GENES FOR SALE

GENES
FOR
SALE

Privatization as a
Conservation Policy

Joseph Henry Vogel

New York   Oxford
OXFORD UNIVERSITY PRESS
1994

Oxford University Press

Oxford   New York   Toronto
Delhi   Bombay   Calcutta   Madras   Karachi
Kuala Lumpur   Singapore   Hong Kong   Tokyo
Nairobi   Dar es Salaam   Cape Town
Melbourne   Auckland   Madrid

and associated companies in
Berlin   Ibadan

Published by Oxford University Press, Inc.
200 Madison Avenue, New York, New York 10016

Oxford is a registered trademark of Oxford University Press

Based on a preliminary version of this work entitled *Privatisation
as a Conservation Policy*, which was published in a special limited
edition for the AIC Conference on Biodiversity, 16– 17 November
1992, Sydney, Australia © 1992 by Joseph Henry Vogel.

Library of Congress Cataloging-in-Publication Data
Vogel, Joseph Henry.
Genes for sale : privatization as a conservation policy /
Joseph Henry Vogel.
p. cm.   Includes bibliographical references and index.
ISBN 0-19-508910-3
1. Germplasm resources—Economic aspects.
2. Biological diversity conservation—Economic aspects.
3. Privatization.   I. Title.
QH75.V63   1994   333.95'16—dc20   93-40210

1 3 5 7 9 8 6 4 2

Printed in the United States of America
on acid-free paper

To Trucanini,
the "Last" Tasmanian

Contrary to expert opinion, extinction is a matter of degree. Trucanini was not the "last" Tasmanian. Aboriginal Tasmanians interbred with European sealers, and their descendants can still be found in several communities along the coastline of Tasmania. What was exterminated was a rich gene pool that diverged from other aboriginal gene pools some 12,000 years ago. Evolutionary theory would predict that within the exterminated gene pool were genes that solved the problems of aboriginal life in Tasmania. These genes were unique and are now extinct.[1]

Trucanini was not just an interesting ensemble of genes. She was also a historical figure who has come to symbolize the plight of native peoples worldwide. Born in 1812 when the first whites settled Tasmania, she witnessed the virtual genocide of her people. Through the cold lens of economic theory, her problem can largely be reduced to one of property rights. She had none.[2] She died in 1876, and even the right over the disposal of her own remains was denied. Although she was buried according to her wishes, her skeleton was subsequently exhumed, packed, and put in storage at the Hobart Museum. In the 1890s, the box labeled "The Last Tasmanian" was unpacked and the skeleton exhibited against earlier government instructions that it never be exhibited.

Over the years, the public increasingly found the skeleton distasteful, and the museum withdrew it from exhibition in 1947. Finally in 1976, Trucanini's right over the disposal of her remains was recognized. However, she was not buried; policing the tomb against theft was deemed too costly. The skeleton was cremated and scattered at sea. In the tortuous fate of Trucanini's remains, we see an economic idea that is central to this book: property rights are never fully delineated.

With the quincentenary of Columbus's "discovery," the North is now celebrating the deprivation of property rights of native peoples like Trucanini. To commemorate, native peoples are calling for a reassignment of existing property rights to address the historical wrongs inflicted on their ancestors and on themselves.[3] I sincerely wish them luck but, at the same time, suggest a more feasible tack: the creation of property rights over genetic information. Not only do native peoples already hold title to genetically rich habitats in their communities and homelands, but they themselves, like Trucanini, embody unique genetic information that may some day yield benefits of commercial value.[4]

Preface

For every complex problem there is a simple solution and it is wrong.
H. L. Mencken

Mass extinction is a complex problem, and its solution will also be complex. In this book, I will attempt to reduce the problem into its component parts and then construct a solution that will invite complexities. The construction derives from an unlikely synthesis: the economics of property rights and the biology of conservation. In explaining the connection of economics and biology, I have been forced to turn to yet another discipline: information theory. Information theory is the lingua franca through which economists and biologists can communicate. Nevertheless, it is not the last word; information theory is tightly linked to a holistic enquiry known as nonequilibrium thermodynamics (NET). NET is the last word. Each of the component parts to the problem of mass extinction can ultimately be traced to NET.

Although mass extinction can be reduced to NET, such a reduction is often not necessary. In fact, it would even divert the reader's attention from the overall thesis of privatization. Therefore, NET will enter the discussion only when the component parts appear unrelated to one another and somehow detached from the thesis of privatization.

Assembling the component parts has often been difficult.[1] For example, there are only crude estimates of the number of species worldwide. Likewise, the rate of species extermination can be inferred only from theoretical models of biogeography. The measurement of economic variables is not much better. Seemingly simple data such as human populations and Third World debt contain shockingly large error terms. This absence of statistics has placed an additional stress on the thesis: the theory must be robust over a wide range of conditions. I believe that no matter what the level of species extinction, human population, or Third World debt, the logic of the policy remains intact.

Because species extinction is complex and complexities do not lend themselves to a simple linear exposition, I have had to separate the logic of the policy from the complex facts that would substantiate it. Fortunately, word-processing programs can accommodate this separation and enable a linear exposition. For these reasons, the substantiating facts are separated into boxes and the explanatory remarks, shunted into notes. If the reader is distracted by these boxes and notes, I suggest just skipping them.

Despite the facts presented, the book is more deductive than inductive. Some may even criticize it as "long on theory, short on facts." I anticipate such criticism and reject it for a very good reason: extinction is irreversible. Inasmuch as the very object of empiricism is being expunged, extinction precludes exhaustive collection, meticulous measurement, and cautious analysis. Extinction demands deduction; the tighter the argument, the more easily the policy can be adopted and implemented.

In teasing out the causality of the mass-extinction crisis, I have committed a breach of academic etiquette: names are named, and culpability is assigned where it is due. The names named are usually those of politicians and economists. The reason for naming names is neither to bash politicians nor to belittle economists, but to persuade the readership of the importance of both. Both were required for the creation of the problem, and both are now needed for the implementation of a solution.

Finally, I must acknowledge that I am not the first to formalize the notion that privatization can become a conservation policy. Roger Sedjo, an economist with Resources for the Future, fleshed out the core idea in an essay entitled "Property Rights and the Protection of Plant Genetic Resources" in 1988.[2] In the same year, sociologist Jack Kloppenburg of the University of Wisconsin suggested similar protection in *First the Seed: The Political Economy of Plant Biotechnology*.[3] The following year, Calestous Juma published *The Gene Hunters: Biotechnology and the Scramble for Seeds*, which traces the emergence of genes as intellectual property.[4] Fortunately, the solution of creating property rights over genetic resources is still sufficiently complex that room was left in the existing literature for my research program.[5] Undoubtedly, there will be other theorists as yet unpublished who are converging on similar deductions. Hopefully, my work will mesh with theirs and help forge a complex solution to the complex problem of mass extinction.

Quito, Ecuador   J. H. V.
May 1994

Acknowledgments

Many readers will be uneasy with the content and style of this book. The book is the expression of a very counterintuitive idea: privatization can become a conservation policy. To air this idea publicly, I have had to abandon the rigid style that characterizes much of economic writing.[1] The prose is simple. Even references have been jettisoned to preserve the flow of the argument. Citations appear in the notes, which serve as an entrée to the technical literature and the notes are followed by a short glossary.

Such a book could not be written without release time from my teaching schedule. Most of this book was written over summers and during a sabbatical leave. From 1990 to 1993, I enjoyed the sponsorship of the Centre for International Research in Communications and Information Technologies (CIRCIT) in Melbourne, Australia. I wish to thank the director, Bill Melody; the deputy director, Don Lamberton; and the dozen social scientists and lawyers on staff. Not only did they listen as I mulled over the implications of privatization, but they also corrected many of my misconceptions.

My good fortune in Australia continued in Brazil with two Fulbright Lecture/Research awards for the autumn semester 1991 and the summer term 1992. The host institution was the Universidade Federal de Viçosa (UFV). I wish to thank Fernando Rocha, head of the Departamento de Economia Rural; Solón J. Guerrero Guttierrez, head of the Núcleo de Estudos e Pesquisas do Meio Ambiente (NEPEMA); and Marco Antônio da Rocha, director of the Fulbright Commission in Brasilia. The awards could not have come at a better time. Mass extinction is a crisis in Brazil, and the only way to contain it seems to be a mass infusion of foreign aid. The United Nations Conference on the Environment and Development, RIO'92, has become the platform from which the money is requested. The Brazilian economy is also in crisis, but here the government looks inward for the solution. It is pinning its hopes on the privatization of state-run enterprises. The proposal of this book seems a logical extension of the current trend. Indeed, if its subtitle, *Privatization as a Conservation Policy*, cannot spark interest in the Brazil of the 1990s, it is certainly not because the timing is bad.

Ecuador is also ready, and I thank Amparo Menéndez-Carrión and Germán Creamer for providing me a new institutional home at the Facultad Latinoamericana de Ciencias Sociales. I am now able to explore the ideas of this book in a small-country context. The challenge is most rewarding. Given the topical nature of *Genes for Sale*, I have not been lacking invitations to lecture in various forums around the world. Due to the excellent feedback from audiences, the policy has been evolving as I have been lecturing. Attendants to any one of the lectures will find a significantly different policy in this book from the one heard in the lecture. Hopefully, the attendants will see their questions more fully answered here. At any rate, I would like to extend my appreciation to those in the following audiences who asked questions and especially those who put me on the spot: CIRCIT (August 1990); the Committee for the Economic Development of Australia (August 1990); the "Biodiversity and Sustainable Development" Symposium of the Seventeenth Pacific Science Congress (Honolulu, June 1991); The New York Botanical Garden (June 1991); A Pontifica Universidade Católica de Campinas (October 1991); Fundação Para a Natureza (October 1991); A Universidade Federal de Cuiabá (October 1991); O Instituto Nacional de Pesquisas Amazônicas (October 1991); O Museo Goeldi (October 1991); Biodiversitás and the Universidade Federal de Minas Gerais (November 1991); O Seminário Internacional de Política Agrícola held at UFV (November 1991); "The Institutional Options for Managing Protected Areas" workshop of the Fourth World Congress on Parks and Protected Areas (Caracas, Venezuela, February 1992); the First Seminar on the Establishment of a State Park Serra do Brigadeiro (Carangola, MG, Brazil, May 1992); the "Optimization of the Environment for Conservation and Development of Tropical Forests" symposium (Forestry, '92, Rio de Janeiro, May 1992); The East–West Center (July 1992); The Department of Economics, University of Hawaii (December 1992); the Masters Programme in International Law, King's College, University of London (February 1993); the Australian Agricultural Economic Society—Victorian Branch (Melbourne, May 1993); The Society for Growing Australian Plants (Cairns, June 1993); Department of Environment Resource Management, Pennsylvania State University (August 1993); the Department of Biology, Hobart and William Smith College (September 1993); the Department of Philosophy and the Department of Forestry, University of New Hampshire (September 1993); the Department of Wildlife and the Department of Resource Economics, University of Maine (September 1993); the Department of Biology, Williams

College (September 1993); the "Conservation Biology Seminar Series," University of Massachusetts, Boston (September 1993); the Intellectual Property Committee, University of Rhode Island (September 1993); the Seminar Series "People and Nature: Views for the 21st Century," Brown University (September 1993); the "Harrison Program on the Future Agenda," University of Maryland (October 1993); the Department of Geography and the Department of Zoology, University of Tennessee (October 1993); the Department of Zoology, Auburn University (October 1993); the Department of Botany, University of Texas, Austin (October 1993); the Department of International Studies, American Graduate School of International Management (November 1993); the Center for Agribiotechnology, Rutgers University (February 1994); the Department of Biology, George Mason University and The Institute for Genomic Research (April 1994); the "Third Annual Conference on Shifting Paradigms in Science and the Environment," College of Environmental Science and Forestry, State University of New York (April 1994); the School of Library and Information Science, Indiana University (April 1994); and the Plant Introduction Station, Agricultural Research Service, United States Department of Agriculture North Central Region (April 1994). An earlier limited edition of the book was published in desktop form for the AIC Conference on Biodiversity (Sydney, November 1992).

Many people also read the manuscript and made contributions according to their expertise. The experts spanned anthropology, biochemistry, chemistry, ecology, economics, education, evolutionary biology, forestry, law, medicine, philosophy, sociology, and taxonomy. I would like to thank Linda Gay Augustine, Michael Balick, Kenneth E. Boulding, Helen Caldicott, Mario Castillo, Mari Davis, Michael DeArmey, Jorge Dergam, David Downes, Newt Fawcett, Larry Hamilton, Allison Hess, Robert F. Housman, Dan Hunter, Gordon Brent Ingram, Lionel Johnson, Ângelo Machado, Allison McGuigan, Jeffrey A. McNeely, Ed Nissan, Walter Parham, Denise Reghenzani, Mario Rizzo, Tim Rogers, R. David Simpson, Supriya Singh, Sy Sohmer, Martin Stokie, Shanti Tangri, and Lionel Tiger. Inasmuch as a synthesis across so many disciplines hazards distortions and mistakes, the usual disclaimer applies.

While working on various drafts of the book, I also enjoyed excellent research assistance: in Australia, from Jenny Connor; in the United States, from Sammond Cheng; and in Brazil, from Moira Adams and Camilo Gomides, who also undertook the translation of the book into Portuguese for the Getúlio Vargas Foundation. I much admire their skill in rendering a translation that is both faithful and flowing. A special thanks also goes out to Adrienne Ralph, John Jannik, Claire Jones, and Henry C. Vogel, who painstakingly proofread the manuscript and zeroed in on its weaknesses. Inasmuch as none of the proofreaders is expert in anthropology and the other disciplines, their proofreading has also proven invaluable. The book will succeed if it is as persuasive to the uninitiated as to those expert in the field.

Contents

1. Introduction, 3

2. Statement of the Problem and Survey of the Solutions, 8

3. Genetically Coded Functions: The Importance
   of Definitions, 15

4. Analogy Becomes Application: Property Rights
   Mean Privatization, 23

5. Illustrating the Problem and the Property-Rights Solution, 32

6. Genesteaders, 40

7. The Rationale, Design, and Implementation of
   the Gargantuan Database, 52

8. Politics, 64

9. Finance, 76

10. Final Payments: Greenhouse-Gas Abatement, 89

11. Conclusion: The Ten Principles for Conserving
    Genetic Information, 103

   Notes, 115

   Glossary, 147

   Index, 151

GENES FOR SALE

1

Introduction

Economics is a rhetorical enterprise. This is the central message of a seminal article that appeared in the economics literature several years ago. Its author, Donald McCloskey, argued that economists persuade not by the weight of the evidence but by the appeal of their words.[1] However, McCloskey does not advocate a "more evidence, better science" approach. Instead, he argues that the scientific method is often unattainable for social phenomena. According to McCloskey, economists should not prostrate themselves over the trappings of science; any rhetorical device, scientific or otherwise, will move the discussion forward.

The urgency of mass extinction lends itself well to the rhetorical approach of McCloskey. Rhetoric is the art of persuasion. Any conservation policy, as a piece of rhetoric, must persuade multiple audiences. Besides the general public in both the North and the South, the audiences include special-interest groups in agriculture, banking, and virtually every industry. The construction of a persuasive policy is all the more daunting when one realizes that the policy must retain its persuasiveness upon translation into different languages for people of different cultures. Under such constraints, simplicity becomes the key. The logic behind the policy must be tight and the progression of the argument, linear. Most important, the policy must be sufficiently adaptable to implement in both the North and the South.

Tight logic and easy flow are not enough. Rhetoric also requires that words be carefully chosen. The word "privatization" in the subtitle of this book unleashes a host of bad images.[2] It seems a poor choice. Nevertheless, "privatization" is the most accurate word to describe the policy I am advocating. To persuade those readers who may harbor prejudices against the *p* word, I will briefly recount the history of its unfortunate connotations and demonstrate how to dissociate the policy I advocate from those connotations.

"Privatization" is immediately associated with the International Monetary Fund (IMF) and its "structural adjustments"—that is, austerity programs. Such programs have induced efficiency by making millions of workers redundant throughout the South. The layoffs occur almost invariably without retraining or any social safety net. The gains in efficiency wrought from privatization are then used to pay interest on debt owed to northern banks. Because the gains are insufficient, other measures are also employed. The most common are deep cuts in social programs. These cuts have translated into malnutrition, higher mortality rates, and a precipitous drop in the overall standard of living.

For these reasons, I was a bit reluctant to use the word "privatization." Many will reject, out of hand, any policy so entitled. Nevertheless, "privatization" is the only word to describe accurately the policy I am proposing. Hopefully, the critics will be sufficiently alert to discern the difference between what I propose and what the IMF mandates. What I propose is the creation of property rights where none existed; what the IMF mandates is the transference of existing property rights from the public sector to the private.

In a nutshell, the proposal is to create legal title over genetic information as it occurs in nature. Such creation will enhance the rents of whoever owns the land. These landowners will enjoy rights analogous to intellectual property rights (e.g., patents, copyrights, trademarks, and trade secrets) whenever the information on their land is commercialized. The first five chapters will explain how conservation is the consequence of such ownership. The remaining chapters will explain how to effect the policy.

The core idea is simple, and simple ideas often fail because of the complexities of the problem. The creation of property rights over genetic information is no different. Rather than ignoring the complexities, I will recognize them even if I am unprepared to deal with them. For example, the question of landownership is beyond the scope of this book, and yet it is fundamentally important to the book. Inasmuch as titles to land are in dispute, titles to genetic information will also be in dispute. Indeed, it is not uncommon for government bureaus, indigenous peoples, and urban investors to have claims to the same piece of land. *Justice in determining who owns what is a necessary condition for the policy.*[3] Where it is determined that the government owns the land, the proposal will advocate a redistribution of public lands to private individuals who will enjoy title to the genetic information. Where it is determined that indigenous peoples own the land, the proposal will imply a windfall gain to these peoples given their low population densities. And where the land is owned by urban investors, the proposal will imply a share tenancy with those who manage the genetic information on the land.

Those who end up managing the genetic information will reap the rewards of ownership. Due to the nature of ecological management, relatively small land tracts will be required for efficient monitoring, preservation, and intervention. Therefore, the policy may even ameliorate the highly skewed pattern of landownership throughout the South. The pioneers of conservation will be a new breed, willing to learn conservation management but no less driven by profit than the traditional frontiersman. In Chapter 6, I call this new breed "genesteaders" and model a feasible reward structure.

Because natural habitats seldom coincide with artificial boundaries, the owner of a tract of land will usually become only one of many co-owners of a piece of genetic information. In other words, many landowners will have claim to the same piece of genetic information. Sorting out each landowner's share is a daunting technical problem. In Chapter 7, I elaborate what I see as the solution.

Ultimately someone must pay for conservation. Who will it be? This is the bottom line. The answer turns on a baseline ethic assumed in the policy: *those who benefit, pay the costs associated with the benefits.* Most people will find this ethic acceptable; its beauty is that it also corresponds to efficiency criteria in economics. In the short run, the corporations (e.g., pharmaceutical, agribusiness, chemical industry) that produce goods that utilize genetic information will pay substantial royalties to the South. In the long run, these royalties will be embedded in the cost structure of industries and will be borne by the consumer. Unwillingness to pay not only violates the baseline ethic, but also means less consumption for the consumer and fewer profits for the corporation. The political obstacles in implementing the policy are discussed in Chapter 8, and the financial ramifications are discussed in Chapter 9.

It would be easy to end the book with Chapter 9. Several reviewers of the manuscript have suggested just that. I have rejected this advice for the simple reason that the policy, as it stands in Chapter 9, does not address an overriding issue in the mass extinction crisis: global warming. It is well known that many species—perhaps millions of species—cannot tolerate an increase in temperature of even a few degrees Celsius per century. To survive in a greenhouse world, species will have to migrate or adapt. Usually there is nowhere to migrate and no time to adapt. Therefore, global warming is not a peripheral issue but a central one to any policy that attempts to preserve genetic information.

In addressing global warming, I am risking the wrath of some very distinguished economists who might otherwise be sympathetic to the policy. Unlike these economists, I do not believe it is my place to discuss whether global warming is fact or fiction. With that said, I quickly add that science can become effective rhetoric for the climatologists who document global warming and against those economists who blithely criticize what(ever) the climatologists document. From the scientific vantage point of nonequilibrium thermodynamics (NET), the greenhouse effect makes almost perfect sense, while the standard economic model makes almost no sense. The importance of NET to the debate will be explained in Chapter 10. If the public can be persuaded that global warming is more likely to be fact than fiction, then it becomes the legitimate role of the economist to discuss the policies that would achieve greenhouse-gas abatement. Among the choices, the privatization of the atmosphere seems the most equitable and efficient. Like the privatization of genetic information, the privatization of the atmosphere turns on another simple unobjectionable ethic: *those who generate a cost, pay that cost*. Chapter 10 is entitled "Final Payments: Greenhouse-Gas Abatement."

Finally, I must remark on the spirit and tone of this book. The dark side of the mass-extinction crisis has instilled in me a certain sense of skepticism, if not cold cynicism. Undoubtedly, this bleeds through. At every step in the construction of the policy, I have asked myself, "How can people circumvent the policy and profit?" I have tried to align selfish interests, but over time changing circumstances will disrupt the alignment. Therefore, conservation is an escalating race between the ingenuity of the policy maker in realigning selfish interests and that of the entrepreneur, who will seek misalignments and exploit them. With this in mind, checks and balances have been incorporated into the policy. Any action based on this policy that guts these checks and balances is doomed to failure and is disowned. Likewise, the ten principles for conserving genetic information in Chapter 11, the conclusion, are not a Chinese menu from which politicians may pick and choose. All ten are necessary to effect the policy.

Is this cynical tone really necessary? Unfortunately, it is. A bit of cynicism can be highly effective in communicating how selfish interests created the problem and how selfish interests can now provide a solution. For many of us in the North, it is a disquieting fact that, for nearly three decades, our banks and governments lent billions of dollars to deforest the South. It is easy to say that we did not know *then* but we know *now*; the ugly truth is that we lent the money over the protests of the environmental community.

Governments in the South also obliged; much was gained by few through deforestation and mass extinction.

The solution is not to replace the selfish in both the North and South with the altruistic. The cynic suspects that, given the same opportunities, the altruistic today could become the selfish tomorrow. This book will suggest that the solution is to channel selfish interests to an end that would be the same result as if there were altruism. The overarching theme of *Genes for Sale* can be stated simply: self-interest can achieve that which inadequate noble intentions cannot. This is hardly new. Recall Adam Smith's most famous lines: "It is not from the benevolence of the butcher, the brewer, or the baker that we expect our dinner, but from their regard to their own interest."[4] This book will demonstrate that Smith's words have never been more timely. Indeed, it will not be from the benevolence of the landowner, the corporate board, or the ecologist that we will expect preservation, but from their regard to their own selfish interests.

2

Statement of the Problem
and Survey of the Solutions

The year 1986 is a turning point in the mass-extinction crisis, not because the crisis was finally contained, but because the crisis was finally publicized. In September 1986, a conference of over 1,000 participants was held in Washington, D.C., under the auspices of the Smithsonian Institution and the National Academy of Sciences. The organizers made no bones about the unscientific intentions of the scientific meeting; the National Forum on BioDiversity was meant to be a media event, and it was indeed a media event. Not only did every major American newspaper cover the story, but journalists reserved the rain forest for feature stories in future issues.[1]

The National Forum on BioDiversity also sparked action. Barely a year later, a grass-roots organization, the New York Rainforest Alliance, had sprung up seemingly from nowhere. The alliance sponsored a three-day symposium at Hunter College in October 1987 with a managerial adroitness that belied its status as grass roots. One year of media blitz could be seen in the sophistication of the audience. The symposium went far beyond the who, what, where, when, and why of mass extinction. Strategies for personal action were explained. At a micro level, individuals were told how they could reduce demand for goods that negatively impacted rain forests: at a macro level, individuals were implored to lobby their representatives. High on the agenda for macro level action was reform of the World Bank lending procedures.

Since 1986, nongovernmental organizations (NGOs) have proliferated. So too have the feature stories on extinction. Inasmuch as the rhetoric of the NGOs seems to be confirmed by the evidence, mass extinction makes good copy. To anchor both the rhetoric and the evidence in the larger scheme

The Antecedent Literature

David Ehrenfeld lays down an ethical code for conservation in *The Arrogance of Humanism* (New York: Oxford University Press, 1978). The philosophy harks back to that of Aldo Leopold, grandfather of the American conservation movement: in short, every species has a right to exist. Lester Brown relates human population to impending disaster in *The Twenty-Ninth Day* (New York: Norton, 1978). Norman Myers documents the extent of the damage in *The Sinking Ark: A New Look at the Problem of Disappearing Species* (New York: Pergamon, 1979), and its economic consequences in *A Wealth of Wild Species: Storehouse for Human Welfare* (Boulder, Col.: Westview Press, 1983). Paul Ehrlich and Ann Ehrlich provide a treatise on extinction: *Extinction: The Causes and Consequences of the Disappearance of Species* (New York: Random House, 1981). And E. O. Wilson articulates the aesthetic losses of extinction in his highly personalized account, *Biophilia* (Cambridge, Mass.: Harvard University Press, 1984). In *The Diversity of Life* (Cambridge, Mass.: Belknap, 1992) Wilson even argues that we must save all species.

of things, journalists usually turn to one of several user-friendly books. These books are written by world-class conservationists and are enjoyable reading in their own right. However, it is doubtful that any one book gets read in its entirety by the journalist covering the story. Some excerpts appear recycled themselves. For example, E. O. Wilson's remark that mass extinction "is the folly our descendants are least likely to forgive" crops up everywhere.

From a strictly rhetorical viewpoint, mass extinction is easy pickings for the journalist. Not only are the references easily intelligible, but the story is also dynamic; each day another species bites the dust. A nice touch to any pending extinction story is the biography of an NGO founder who came to the rescue of a pink dolphin or a spotted owl. The biography is usually put in a box separate from the story. Inside the box, one learns that the founder is not a specialist but just your average guy, perhaps scruffier than most, whose exceptional quality is "internal locus of control"—that is, the belief one can make a difference. The story ends with a glimmer of hope. Fishnets that entrap the pink dolphin are banned or the permits to log old growth forests are rescinded. What more could any journalist want?

For the purposes of policy analysis, the problem of extinction needs no fuller statement than that provided in the aforementioned books or even the current news stories. Therefore, I see no need to recite the best estimates of extinction. Others have done this well. What I will recite is the rhetoric surrounding the statistics and explore its economic impact. The rhetoric as well as the statistics have helped shape the solutions. These solutions will serve as a standard against which privatization will be evaluated and, if successively argued, embedded.

Conservationists were around and kicking long before the crisis hit the newsstands. Only recently, however, have they realized that the public is not persuaded by bad news presented in tables and charts. Michael Soulé, one of the architects of modern conservation biology, makes this point wryly: "Physiologically, bad news is depressing, and depression inhibits arousal in the limbic-emotional system. Advertisers and politicians know this tacitly: consumers don't buy coffins, even when on sale, and voters don't elect prophets of doom. Perhaps this is one reason why there are no biologists in the U.S. Congress."[2]

One should not expect the public to be persuaded by numbers. The estimate that 2 million species will exit Earth is almost the equivalent discomfort level as an estimate of 20 million. Even an order of magnitude goes over the heads of those who must be persuaded: the voters in a democratic nation. Realizing this, the conservationists have changed tactics and launched an advertising campaign that would put Madison Avenue to shame. The ploys are so obviously psychological that, in mixed scientific company, they should embarrass the conservationists who deploy them.[3]

Just as Madison Avenue can sell junk if properly packaged, even bad news can elicit action if delivered well. It is no accident that the media campaign links deforestation and mass extinction to food, home, and family. Once the image is stuck in the emotive recesses of the brain, the rain forest becomes a highly personal issue. To know that 50 percent of all species will go extinct by the year 2000 is less powerful than establishing the causal relationship between you, the reader, and the cumulative threat to your food, your home, your family.

Let me offer my favorite image to emerge out of the rhetoric: the ground-meat connection. Over the last few years, advertisements appeared in the North adopting the following format: Each _____ hamburger (fill in the blank according to which retailer is still buying tropical beef; for example, McDonald's, Burger King) destroys an area of rain forest the size of your kitchen, and this gluttony summed across individuals destroys *each year* an area of rain forest the size of _____ (fill in the blank according to the geographic location of the target audience; for example, West Virginia in the United States, the United Kingdom in Europe). Multiply the number of hamburgers you eat each year by your kitchen floor, and this is your crime against nature. Guilt now mingles with the simple pleasures of fast food.

Ground meat is not enough. So there are other highly effective rhetorical tools. The medicinal values of endangered plants are often exemplified through two chemotherapeutic drugs: vincristine and vinblastine. The active ingredient of both derives from the rosy periwinkle, a species native to Madagascar. The revenues from both drugs have passed the $1 billion mark. I do not wish to belittle the value of the rosy periwinkle in terms of either dollars or human lives. However, the overemphasis on the periwinkle is embarrassingly manipulative. Who in the North has not lost a relative through cancer? The rhetorician within the conservationist reasons: link the demise of the rain forest to the demise of a relative, and consolation and duty become intertwined. Enter again vincristine and vinblastine in every feature story on mass extinction.

I suspect that the sensationalization of hamburgers and kitchens, chemotherapies and rosy periwinkles, is risking a satirical backlash. To date, conservation has escaped the wit of *Punch* in the United Kingdom or *National Lampoon* in the United States. Nevertheless, the makings of a backlash are in plain view. For example, Brian J. Huntley, a South African conservationist, laments:

> Biotic diversity is not linked to the distribution of elephants, rhinos, and other so-called charismatic megaherbivores. The massive investment in conservation campaigns directed at these species does more for the souls of the donors and the egos of the elephant experts than it does for biotic diversity, which is centered on less exciting communities of montane forests, Mediterranean heathlands, wetlands, lakes, and rivers.[4]

Huntley's remarks are worthy of mainstream economic thought; rationality would dictate a simple reallocation: fewer elephant experts and more wetland experts. The only problem with such reasoning is that rational economic man does not exist. People respond better to the charismatic megaherbivores. Nevertheless, Huntley's point needs reiteration: greater protection can be had from rerouting the last thousand dollars from the millions spent on the elephants to, say, alkalizing the acidified wetland of an endangered batrachian. But again the reality: the public will not fork over the money for a frog in the swamp! Diverting donor dollars to less appealing species will mean fewer dollars donated.

Unlike Huntley, most conservationists believe that Madison Avenue is the way to go. In taking that road, they are paying respect to the power of

Dissecting *Biodiversity*

The contributors suggest five broad policies to address mass extinction. Below is a list of contributors indexed by policy emphasis.* Most advocate a mix of policies and are only cited in the area of emphasis. In some instances, however, the contributor's position on the other policies is one of vehement opposition. For example, David Ehrenfeld believes that cost–benefit analysis will actually promote extinction.

Education	Regulation	Government Subsidy	Cost–Benefit Analysis	Technology
Brown	Goodland	Cairns	Brady	Altieri and Merrick
Ehrenfeld		Lovejoy	Hanneman	Burley
Ehrlich				Randall
				Ocana
				Robinson
				Shen
				Spears

Note:

Education refers to public awareness promotion, the inculcation of a conservation ethic, and population control programs.

Regulation refers to governmental review of projects on environmental criteria. Such review is either direct, as in the case of public roads, or indirect, as in the case of lending institutions.

Government subsidies refers to the subsidization of commercial activities that preserve biodiversity or the direct acquisition of habitats.

Cost–benefit analysis refers to improvement in accounting of external costs of habitat degradation as well as external benefits of habitat preservation.

Technology refers to the implementation of appropriate techniques to enhance productivity of land use and relieve stress on habitats.

*E. O. Wilson ed., *Biodiversity* (Washington, D.C.: National Academy Press, 1988).

consumer preferences even as those preferences interfere with optimal allocations. Not only do they realize that they must tolerate misallocations, but they also realize that the obvious solution, or what seems to them to be the obvious solution, would make for very bad rhetoric. That unspoken solution is large-scale government intervention.[5]

A caveat is in order. I do not wish to give the impression that solutions other than government intervention have not been tendered. Many can be found in *Biodiversity*, a thick volume of articles collected from the National Forum on BioDiversity and published in 1988. In dissecting the policy mix of *Biodiversity*, I was able to classify five types of solutions. Among them, government subsidies is the sleeping giant. It portends large-scale intervention. Amazingly, no estimate of government subsidies is attempted, even though there are pages upon pages of what will be lost if we fail to preserve species—the litany of vincristines and vinblastines.

Biodiversity is not just a book stating the problem and surveying the solutions; it also is a reflection of attitudes toward government. Because almost each contributor sees government intervention as the unspeakable final solution to the mass-extinction crisis, each puts a heavy emphasis on education. The plan of attack is terribly civilized. It can be summed up nicely as "Sensitize and Sacrifice." "Sensitize" is the euphemism for inculcate. "Sacrifice" is the euphemism for tax. The thinking seems to be that once the public is sensitized, the bitter pill of sacrifice will become palatable. So many of the contributors may not even be heartened that partial solutions are tendered in *Biodiversity*; the final solution, large-scale intervention, is still in the offing.

There is no better testimony to this sentiment than the words of the editor, E. O. Wilson. At the outset of *Biodiversity*, Wilson expresses a resignation that is widely shared: "In the end, I suspect it will all come down to a decision of ethics."[6] Such sentiment departs radically from the feisty position Wilson took just four years earlier in *Biophilia*: "The only way to make a conservation ethic work is to ground it in ultimately selfish reasoning. . . . An essential component of this formula is the principle that people will conserve land and species fiercely if they foresee a material gain for themselves, their kin, and their tribe."[7] In *Biophilia*, Wilson even went so far as to assert that this generation owes nothing to the next generation and that economics should dictate conservation!

This book is more in keeping with the Wilson of *Biophilia* than the Wilson of *Biodiversity*. It will explore the mechanism of property rights as a viable and robust solution. To most conservationists, the idea may at first sound a bit silly. Conventional wisdom dictates that government should acquire property in order to preserve it and that bureaus be expanded in order to manage the new acquisitions. In the chapters to follow, I will suggest just the opposite. I will even be so bold as to suggest that privatization is not just another solution for the box; it is a solution more sweeping than all the other solutions combined. This is an ambitious assertion. It requires that I state precisely what will be privatized. A species? a race? an organism? And who will own whatever it is that is privatized? The state? the corporation? the individual? The language of biology becomes a Procrustean bed for such discussion. Therefore, I see no alternative but to redefine some key biological concepts and cast the discussion in a language more amenable to privatization.

3

Genetically Coded Functions:
The Importance of Definitions

There are many metaphors for evolution. Darwin saw it as a branching bush, and the bush metaphor has enthralled evolutionists ever since. It seems a precise metaphor. Nevertheless, no metaphor is absolutely precise. All metaphors somehow constrict the phenomenon they attempt to illustrate. The branching bush is no different, it is just more subtle. The constriction is felt in economics. To the economist, evolution as a branching bush implies nothing for the cost–benefit analysis of development projects in the tropics.[1] And the bush is not alone. Other fundamental terms in biology also imply nothing. In this chapter, I will introduce a term that obviates cost–benefit analysis and enables discussion of a market for the products of evolution. In the next chapter, the economic rationale for creating this market will be elaborated. But before racing ahead, I must first explain the terminology of biology and how that terminology fails economists.

What is evolution? What is biodiversity? What is extinction? These are questions seldom asked, much less answered, in the conservation literature. Conservationists believe there is no need to construct each argument from the ground up; some concepts are assumed to be common knowledge. However, there are small cracks in the foundations of biology that become fault lines in the complex discussion of conservation. The cracks are in the very terms "evolution," "biodiversity," and "extinction." To appreciate the fault line, one begins with the question least often asked: What is evolution?

In the modern neo-Darwinian synthesis, evolution is defined as a change in gene frequencies. Over time, new genes appear through the process of mutation and existing genes disappear through the process of selection. There

15

is no balance between the entry and exit of genes. On net, the slow accumulation of genes implies progress in the history of life. This progress does not occur continuously, nor is it unidirectional: there have been periods of time when the total number of genes has decreased. The decrease may occur due to an ecological catastrophe or to the intense selective pressures of a few colonizing species. Nevertheless, over geological time—hundreds of millions of years—habitats rebound from catastrophes and the total number of genes increases. Today we are witnessing one of those periods of time in which the total number of genes is decreasing at a rate probably unmatched since the age of dinosaurs, 65 million years ago. The decrease today is almost wholly the consequence of one colonizing species, man.

To interpret evolution as increases or decreases in the total number of unique genes is to interpret evolution in terms more amenable to economics. Discussion in these terms would be a step in the right direction. However, there is no need to stop at the level of the gene. From biochemistry, we know that genes themselves are nothing more than a sequence of purine and pyrimidine bases bonded on a backbone of phosphate sugar molecules. As a sequence, genes can theoretically be assigned probabilities of occurrence. These probabilities can in turn be quantified as bits of information by the Boltzmann equation of thermodynamics or the equivalent Shannon–Weaver equation of information theory.[2]

To reduce evolution to changes in bits of information is to take the high road in scientific reduction. It is also beyond the state of the art. So this line of argument seems to take us nowhere—a detour from the overall thesis of conservation. Surprisingly, it is not. Evolution is about the flow and accumulation of information. Conservation is about the retention of information. However, conservation is not concerned with the retention of information per se; it is concerned with the functions coded in that retained information. Retention is desirable because information codes for functions.

All this sounds very abstract. A simple nonbiological example can illustrate the abstraction. Consider the word "help." By the Shannon–Weaver equation, the word "help" has the same amount of information as the sequence of letters "pleh" ("help" spelled backward). However, the word "help" has meaning; the sequence "pleh" does not. The point is that some information codes for function—"help"; much codes for gibberish—"pleh." Hence, the earlier claim: conservation is not concerned with the retention of information per se; it is concerned with the functions coded in that retained information—only the meaningful sequences.[3]

The above analogy cannot be refused in the realm of biology. The Japanese geneticist Motoo Kimura has spent a lifetime establishing the fact that many genes code for nothing. That is to say, much genetic information has no function. Kimura calls the gibberish genes, "neutral" and the change in the frequencies of gibberish genes, "neutral evolution."[4] The best example of neutral genes can be found in the salamander genome. Salamanders have several times as much genetic information as have humans. It is not anthropocentric to protest that humans are far more complex than salamanders even though humans have far fewer bits of information. Humans are *functionally* more complex than salamanders, but *structurally* they are far simpler.

What is important is not the probabilities of the bases—the genetic information—but the functions coded by these bases. This means that the genes common to species, populations, and even individual organisms should be redefined in terms of functions. These functions are intimately tied to genetic information; however, as Kimura has established, not all genetic information is tied to function. Much is neutral—the gibberish genes. Neutrality means that the object of conservation should be the genetically coded functions. In fact, the phrase "genetically coded function" is the linchpin to privatization as a conservation policy. Because the term is so clumsy, I will refer to it simply as GCF. Once we start thinking of evolution as changes in GCFs, implications emerge for the economist. Before elaborating them, we must first carefully consider the term that GCF will replace—biodiversity.

What is biodiversity? Above all, "biodiversity" is a catchy word. As such it makes good rhetoric. As we saw in Chapter 2, the word "biodiversity" graces the handsome cover of the collected volume of articles from the National Forum on BioDiversity. In the introductory remarks to the book *Biodiversity*, Wilson identifies the father of the term without identifying the child: "Dr. Rosen represented the NRC/NAS (National Research Council/ National Academy of Sciences) throughout the planning stages of the project. Furthermore, he introduced the term *biodiversity*, which aptly represents, as well as any term can, the vast array of topics and perspectives covered during the Washington forum."[5]

In one respect Wilson is right. The term has passed the test of popularity in the marketplace of rhetoric. It has gained currency worldwide and will soon find its way into dictionaries. However, Wilson is wrong in the marketplace of science. The term "biodiversity" has yet to be tested for meaningfulness. In science, that test is formal logic. Criteria in formal logic dictate that definitions be broad enough to capture the essential attributes of the thing defined yet narrow enough to discriminate.[6] One can easily show that by the criteria of logic—breadth and discrimination—the term "biodiversity" must be rejected.

With Respect to Breadth

Most people interpret biodiversity as species diversity. This misrepresentation is not lost by some of the contributors to Wilson's *Biodiversity*. G. Carleton Ray puts this plainly: "It surely is not merely species variety, as some of the public may be led to believe."[7] Logic demands that neologisms be defined. If "biodiversity" is left undefined, as Wilson leaves it in the quotation above, then people will assign their own meaning to it. Usually, that meaning will be species variety. As such, the term would exclude races (subspecies). However, there is great deal of wealth in races, as any breeder will attest. So we might want to include races. But why stop at races? If races are to be included, why not individuals? There can also be a great deal of value in individuals. For example, the tissue taken from the recovered leukemia patient John Moore is worth billions of dollars.[8] But if we include individuals, then "biodiversity" means the same thing as "biota" and there is no need for a new word.

With Respect to Discrimination

The word "biodiversity" provides no mechanism to discriminate the twin attributes of the biota essential for conservation: function and uniqueness. Each is necessary, and neither is sufficient. For example, John Moore, you, and I are all unique constellations of genes, and as such we are all endangered pieces of the biota.[9] But uniqueness is not enough. Only John Moore had the tissue that met the dual criteria of conservation: uniqueness and function. You and I, in all likelihood, do not.

Unlike "biodiversity," "genetically coded function" (GCF) discriminates the essential attributes of the biota needed for conservation. The discrimination is by function irrespective of the taxon in which the function occurs, as is shown in the box. From the summary, we see that one of the most valuable GCFs for the rain forest is the waste-recycling task of ants. This GCF occurs at the taxon of the family, with its some 8,800 known species. Even though waste recycling is priceless, no one ant species should be preserved because of it; there are 8,799 other species providing, to some degree, the same GCF. Only if we exterminate ant species by the thousands will function couple with uniqueness and warrant the conservation of ants species.[10]

The ant argument is less true for bears. The paws and bladders of bears are valuable GCFs throughout Asia. But like waste recycling among the genera of the family Formicidae, Asians will still enjoy ursine delicacies, to

Functions at All Levels: Rethinking Taxonomy

Family (Formicidae)
Of the known 8,800 ant species in the family Formicidae, all provide the GCF of recycling natural wastes. Inasmuch as forestry would be impossible without the ants, their commercial value is incalculable. It is no hyperbole that the survival of the forests depends on the GCFs of ants.

Genus (*Ursus*)
Six of the eight bear species in the genus *Ursus* are being poached. Bear paws are considered a delicacy in Korea and Japan and fetch as much as $800 per plate; gall bladders are an aphrodisiac and command $50 per gram. Both paw and bladder can be considered GCFs.

Species (*Zea diploperennis*)
In 1979 a wild species of maize, *Zea diploperennis*, was discovered in Mexico. Unlike domesticated maize, the wild species is perennial and highly viral resistant. Two economists estimate the value of perennity (just one of the GCFs!) at $6.82 billion.

Subspecies (*Solanum melongena*, 418 landraces)
Specimens of the 418 landraces of the eggplant *Solanum melongena* are stored at the Southern Regional Plant Introduction Station of the U.S. Department of Agriculture in Griffin, Georgia. A few of these landraces contain genes that code for resistance to *Verticillium* wilt. In 1977, the dollar value of each hectare of eggplants lost to pests was $630.

Individuals (John Moore)
John Moore is a recovered leukemia patient who had an interesting spleen. The diseased cells of the spleen were harvested and grown in the laboratory. The resultant cell line was patented as "MO" and has multiple uses for combating immune deficiencies. The market value is estimated at $3 billion.

some degree, as long as every one of the eight species of the genus *Ursus* is not exterminated. On the basis of selfish interests, each of the eight bear species should be protected in inverse proportion to the substitutability of the paws and bladders of one species with the paws and bladders of another.[11]

As we rethink taxonomy, the same logic holds. The reason for protection at the species level is uniqueness of a GCF at that level. Probably the best example is maize. Over the past century, there have been exhaustive searches for wild species of maize. Almost always the searches have been to no avail.

So it is fairly safe to say that the GCFs of the recently discovered *Zea diploperennis* are unique at the level of species.[12]

The identification of what to conserve is obvious once a GCF has been discovered. But before discovery, one does not know which bits of endangered genetic information will some day code for a valuable function. Therefore, a strategy must be implemented before extinction rules out the possibility of discovery. This means that discrimination of the biota must take place under uncertainty.

Fortunately, the uncertainty is not absolute. We do know that there is a correspondence between uniqueness and the taxon at which the GCF is found.[13] For example, one can say that the likelihood of a GCF unique to an individual is slim (albeit there will be some John Moores). It seems reasonable a priori that no specific organism be conserved. Or to put it in the rhetoric of formal economics, the expected gain from preservation is less than the cost. However, if one discovers that genes code for function at the level of the individual, then uncertainty is removed and the genes should be preserved. A fortiori the gain outstrips the cost. This was the case with John Moore.

As a general rule, GCFs at the level of the individual are most likely shared by other individuals within the race, so they will take lowest priority for conservation. Individuals are substitutable. Much higher priority is given to protection of individuals within an endangered race. On average, these individuals are less substitutable with individuals from different races. Even higher priority is given to individuals within an endangered species. On average, these individuals are much less substitutable with individuals from different species but the same genus. And still higher priority is given to individuals within an endangered species that has no other species in its genus. On average, these individuals are much less substitutable with individuals from other genera but the same family. The highest priority will be given to individuals within an endangered species that is the last of its genus that is the last of its family . . . Most likely, an organism that is the sole occupant at the taxon of family or order will have many GCFs not found in other organisms from different families or different orders.

The reader may begin to see how rephrasing evolution and biodiversity in terms of GCFs has direct implications for conservation. If not yet clear, it will become obvious when the last of the three questions posed earlier is answered: What is extinction? In the terminology of biology, extinction is abrupt: it is the loss of a species. In the model of privatization, *extinction is a matter of degree*: it is the loss of GCFs that can occur at any taxon—from the individual down to the order. On an expectation basis, the loss of the last species of an order is the multiple loss of GCFs, while the loss of a specific individual within a healthy population is a loss of GCFs so negligible that it can be ignored.

By the dual criteria of formal logic, breadth and discrimination, the term "GCF" is superior to the word "biodiversity" (however that word is defined) and should replace it. But wait, there is still the matter of rhetoric. As claimed at the outset of this book, rhetoric as well as science shapes conservation policy. Therefore, conservationists must be persuaded that the survival of the term "biodiversity," even as a piece of rhetoric, bodes ill for conservation policy.

I offer two examples. Neither is hypothetical. Imagine a nation swayed by nongovernmental organizations (NGOs) that promote biodiversity. To the public, biodiversity means species. The nation rallies behind conservationists and legislates endangered species laws that give equal protection to all the "higher" species. However, protection costs money. Therefore, in the spirit of the law, as much money is allocated to protect species A (which is one species among a hundred in its genus) as species Z (which is the last of its order). Implicit in the "save species" legislation was an equal weighting of all species A to Z. Hence, the term "biodiversity" distorts sound conservation objectives.

The second example extends the perversion of the first. Imagine a nation also swayed by NGOs that promote biodiversity. But here the public interprets biodiversity as the biota—all things great and small. The nation becomes outraged to learn that a *decline* in local biodiversity is just what the ecologist orders: the elimination of one species to increase the range of another. To the horror of ecologists, we are starting to see this in Australia. Animal-rights activists are grumbling over plans to eradicate the fox.[14] In Australia, the fox is an introduced species and has no natural predators. Its numbers have exploded. One of the casualties of the fox is the duckbill platypus. However, the platypus is not just another species. It is one of only three species left in its order, the monotremes. On an expectation basis, the platypus has many GCFs that are unique. And on an expectation basis, the foxes in Australia have no GCFs that are unique.

Although the term "GCF "can help the economist rank the conservation value of organisms like foxes and platypuses, the term still cannot assist him, in any meaningful way, in quantifying those benefits! The impossibility of quantifying the benefits is made brilliantly clear by the conservation biologist David Ehrenfeld:

> There are two practical problems with assigning value to biological diversity. The first is a problem for economists: it is not possible to figure out the true economic value of any piece of biological diversity, let alone the value

of diversity in the aggregate. We do not know enough about any gene, species, or ecosystem to be able to calculate its ecological and economic worth in the larger scheme of things. . . . I cannot help thinking that when we finish assigning values to biological diversity, we will find that we don't have very much biological diversity left.[15]

The purpose of coining the term "GCF" is not to assign values to biological diversity and assist cost–benefit analysis. The purpose is to obviate the need for cost–benefit analysis. The term "GCF" enables discussion of a market for the product of evolution: genetic information. By creating a property right over that information, the market will bid up the value of habitats on the likelihood of GCFs far above the average value of alternative land uses. To understand how to create such a market, one must deploy the economic theory of property rights.

4

Analogy Becomes Application:
Property Rights Mean Privatization

Foreign languages are tricky. Quite often, identical words have virtually identical meanings (e.g., "organisation" in both French and British English). Such words are called cognates, and they facilitate the learning of a foreign language. However, not all identical words are cognates. Many identical words can mean something slightly different to something completely different and, on occasion, even something obscene. French teachers call such words *faux amis*, or "false friends."

Faux amis happen not only across languages, but also within the same language. Take, for example, the word "football." Football in the United Kingdom is not the football of the United States, which is not the football of Australia. As a *faux ami*, the word "football" is insidious because the meanings lie in a gray area. All the footballs are related: all are played with feet and balls, but, nevertheless, all are quite distinct; an American would distinguish them respectively as soccer, football, and Aussie-rules football. It is far easier for an American and an Australian to confound the meaning of a word like "football" than to confound the meaning of a word that, by its context, clearly means something different.

A hypothetical situation can illustrate the subtlety or bluntness of a *faux ami*. Imagine an American and an Australian football enthusiast engaged in conversation. Both may think, at least for a while, that they are talking about the same game. But when the American asks the Australian what team he "roots for," the Australian will suspect, by nature of the context, that they are now talking about something very different: "rooting" in America means cheering, while in Australia it means fornication. Whereas "rooting" flags a *faux ami,* "football" does not.

The *faux amis* quality of the term "property rights" is more like that of "football" than "rooting." This means that a lawyer and an economist can easily engage in a conversation about property rights without realizing they are talking about two different things. Each may dismiss the other's valid argument as invalid simply because both equivocate on the meaning of the term "property right."

To the lawyer, "property right" is a very generic term. It refers to any type of right to specific property. This raises the question: What is a right? According to *Black's Law Dictionary,* "a *right* is well defined as a capacity residing in one man of controlling, with the assent and assistance of the state, the actions of others."

To the economist, the term "property right" is much more fluid than the legal definition. When economists talk property rights, they are talking about who controls production processes, the disposal of resources, and the ability to derive income from those processes and resources; those who control, dispose, or derive a benefit have some property right.[1] Neither "assent and assistance of the state" nor "the actions of others" is necessary or sufficient for having a "property right" in economics.

There is another way of explaining the distinction between the two meanings of "property right." The legal and economic meanings can be reduced to notions of de jure and de facto. According to *Black's Law Dictionary,* de facto "is used to characterize an officer, a government, a past action, or a state of affairs which must be accepted for all practical purposes, but is illegal or illegitimate. In this sense it is the contrary of de jure, which means rightful, legitimate, just, or constitutional." When lawyers talk rights, they talk about rights that would be recognized in a court of law, that is, they talk de jure. The rights are discrete: either they exist or they do not exist.

To the economist, rights are continuous and no line is drawn between de jure and de facto. What separates a right de jure from a right de facto is the magnitude of the costs. A right de facto usually means a higher cost for the individual in the control of property, its disposal, and the derivation of value than if the right were de jure. The reason for the higher cost is simple. When a property right is de jure, the executive and judicial branches may assist in the control, disposal, and derivation of a benefit, when a property right is de facto, the executive and judicial branches may resist.

I do not wish to extend the definitional discourse of the last chapter to the term "property rights." The intent is to alert the reader to the differences and lay a foundation for a clear understanding of the economic meaning of

Simple Word Substitutions

The dynamic nature of property rights was first fleshed out in the economic literature by Harold Demsetz:

> Consider the problems of copyright and patents. If a new idea is freely appropriable by all, if there exist communal rights to new ideas, incentives for developing such ideas will be lacking. The benefits derivable from these ideas will not be concentrated on their originators. If we extend some degree of private rights to the originators, these ideas will come forth at a more rapid pace.*

Now substitute a few words into the above quotation:

> Consider the problems of biodiversity and mass extinction. If genetic information is freely appropriable by all, if there exist communal rights to genetic information, incentives for preserving genetic information will be lacking. The benefits derivable from these genes will not be concentrated on those who control habitats. If we extend some degree of private rights to those who control habitats, the genetic information will be preserved at a more rapid pace.

*Harold Demsetz, "Toward a Theory of Property Rights," *American Economic Review*, no. 2 (1967): 359.

the term. Such a foundation is prerequisite to understanding fully the rationale for ownership over genetic information. From this point on, unless otherwise stated, a property right will refer to the meaning assigned by economists.

The economic meaning requires some elaboration. It relies heavily on a comprehensive accounting of costs and the assumption of individual maximization. In other words, the economic approach to property rights assumes that both the producer and the consumer will act to maximize their own pleasure, given a full accounting of the costs that impinge on them. Therefore, the delineation of who controls, disposes, and derives benefits from a piece of property turns on one simple criterion: Is it worthwhile for the individual?

A simple deduction emerges from the assumption of maximization: when the cost of delineating a property right is greater than the benefit of the delineation, the property right is abandoned. This deduction is at the heart of the property-rights approach and examples abound where delineation is too costly and rights are abandoned. The favorite in the literature seems to be seats in a movie house. As any movie goer knows, not all seats are created equal. Some are better than others. By microeconomic theory, each seat should be priced differently according to demand conditions for that seat. That is to say, a seat in the center should fetch a higher price than a seat in the front row near the left wall. Yet in the real world, all seats are priced equally. Why? The economic answer: the marginal costs of policing the seating of patrons is greater than the marginal revenue to be gained from pricing seats differently. In plain English: it costs more to hire ushers than the owner would earn from selling the better seats at a premium. Hence, the

---

From Pelts to Genes

Harold Demsetz illustrated the emergence of property rights with the ownership of beaver pelts by the Montagnais Amerindians of seventeenth-century Quebec. With the advent of the French fur trade, beaver dams and the pelts inside became the property of particular tribes. Demsetz argued that the reason property rights emerged among the Amerindians of Quebec but not among the Amerindians of the plains is habitat size: there were "no plains animals of commercial importance comparable to the fur-bearing animals of the forest, at least not until cattle arrived with Europeans. The second factor is that animals of the plains are primarily grazing species whose habit is to wander over wide tracts of lands."* In other words, the transactions costs of delineating the plains (enclosure of the commons) are much greater than those associated with delineating beaver dams. In the plains, the attribute "grazing species" was abandoned; in Quebec, the attribute "beaver pelts" was delineated.

Through the advent of twentieth-century biotechnology, the demand for genes has risen suddenly, not unlike the sudden rise in the demand for fur pelts in seventeenth-century Quebec. To the extent that the valuable genetic information is distributed across taxa and land titles, the delineation of the genetic-information commons will be like that of the grazing species of the plains: difficult. To the extent that the genetic information is concentrated in small habitats, the delineation will be like that of beaver dams: easy.†

*Harold Demsetz, "Toward a Theory of Property Rights," *American Economic Review*, no. 2 (1967): 357.

†Demsetz foresaw that property rights would emerge as solutions to many new externality problems: "A rigorous test of this assertion will require extensive and detailed empirical work" (350). What Demsetz did not foresee was the irreversibility of the externalities. Because genetically coded functions can be irreversibly lost in the hit-and-miss emergence of property rights, the institutionalization of property rights becomes the proper role of government.

owner throws the seats out into the public domain—the property right is abandoned.

Once a property right is abandoned, individual maximization will again determine how the resource is allocated. In this example, the patrons decide if the value of a good seat is worth arriving early. In other words, it is in the patron's interest to be first in line if the value of his time in line is less than the premium he would pay for a good seat. He who lines up early becomes the residual claimant of the attribute "good seat location."

In this example, the high costs of policing seats make the early patron the residual claimant. But different goods will have different relative costs. For some goods, the costs will be high in policing; for others, in monitoring or negotiating. Any cost associated with the transfer, capture, and protection of a right is called a transaction cost. To delineate a property right, one chooses a method that minimizes the transaction costs and maximizes the capture of benefits.

Over time, both the transaction costs and the benefits will change. Changes in technology and changes in demand conditions can change the calculus of delineation and the decision to abandon a property right. Hence, the property-rights approach is a dynamic model of economic organization.

One can extend the movie-house example to illustrate the delineation triggered by changes in technology and changes in demand. Imagine that an entrepreneur invents an inexpensive keycard-latch mechanism on seats. The patron buys a keycard rather than a ticket, which entitles him or her to sit only in a designated seat. The card is inserted into a designated locked upright seat, and the seat drops. Such an invention would enable exclusion and capture the value of seat location. If the premia on good seats are greater than the installation and operation costs of the keycard mechanism, "good seat location" will be delineated. Once delineated, the residual claimant will shift from the early-bird movie goer to the movie-house owner.

The impact of a change in demand is similar to that of a change in technology. Imagine that the movie house converts into a theater. Theater shows fetch higher prices than do movies. The theater owner will find it profitable to number seats, hire ushers, and discriminate seat prices. The premia on movie-house seats that were left in the public domain will now be captured simply because those premia have risen. In other words, it now proves profitable for the owner to further delineate a property right over seat location.

This line of reasoning can easily lead one up the garden path. One might infer that given sufficient changes in exclusionary mechanisms or demand, property rights will, at some point, become fully delineated. To understand why this is wrong, consider again the movie house-turned-theater. It is plausible that during intermission, an occasional patron in the mezzanine will move to an empty seat in the orchestra. To delineate fully a property right over seat location would require policing costs for the occasional usurper. No doubt it will be cheaper to allow the occasional seat to the usurped than to police for this contingency. A strong implication emerges from the continuous nature of transaction costs: *it is never worthwhile to delineate a property right fully.*

The notion that property is delimited by costs and especially by policing costs may leave the reader uneasy; it seems so cynical. Nevertheless, it is realistic and goes far beyond the small-scale example of a movie house and seat-usurping patrons. Large-scale examples abound in both the North and the South. I will offer two of broad interest to conservationists. In the United States, the Reagan–Bush administration has been accused of abusing its budgetary power over the Environmental Protection Agency (EPA) to reduce enforcement of laws that it did not like and that it could not change through the legislature. The case can be made that under weak EPA enforcement, polluters enjoyed, to some degree, a de facto property right to pollute that was denied them under the Carter administration. The same phenomenon, but on a grotesque scale, has occurred in post-colonial Africa. Since independence, poachers and squatters have converted nature reserves into what can more accurately be described as "paper parks."[2] With little, if any, public money spent on policing these reserves, poachers and squatters are enjoying a de facto property right over them.

Given knowledge of the transaction costs, the property-rights approach can predict resource allocation (such as the paper parks in Africa) and explain organization (such as the ushers in the movie house-turned-theater). Explanation and prediction are the sine qua non of good science. However, as argued at the outset of this book, economics is not a science. It is a rhetorical enterprise. Resources are not allocated on the basis of logic. Resources are allocated on the basis of persuasion. The scientific method is just one tool of persuasion, and economists seldom apply it correctly.[3] What is persuasive is practicality. Indeed, practicality should be the sine qua non of economics. The best economic theories are those that can be packaged into comprehensible decision-making tools.

On practicality, the property-rights school can fare well. A practical implication leaps out of transaction-costs calculus: if transaction costs can be reduced, efficiency will increase. The reason: trade. Values that were abandoned under high transaction costs are captured as transaction costs are reduced and property rights are delineated. With delineation, the attributes will be traded to those who derive the greatest value from them. Think again of the movie house and the keycard mechanism. Once the property right is delineated over seat location, the owner earns more money from the patrons who most value good seats. No one wastes a minute in line. In sum, resources are allocated more efficiently.

The property-rights approach is the theoretical basis of privatization as a conservation policy. *A property right over genetic information will enable trade and increase efficiency as long as the value of the genetic information traded is greater than the associated transaction costs and the value of alternative land uses.* To appreciate the associated transaction costs of conservation, a more relevant analogy is needed. After all, seat location in a movie house is a very different asset from the genetic information in a habitat. Fortunately, there is an example in the property-rights literature that seems tailor made to an analogy with genetic information: *copyrights over computer software programs* is to *the creation of software* what *the privatization of genetic information* is to *the conservation of genetically coded functions, GCFs*.[4]

To see the analogy, let us first consider the software writer and the software. Imagine a writer in a world where there is no property right over software. The writer who creates new software has generated a benefit that can be enjoyed almost costlessly by anyone with a computer and the master diskette. In such a world, the only consumer who will buy the program is a consumer whose application is worth more than the cost of creating the software. In other words, the consumer will have to compensate the writer for the months and even years spent in writing the program. However, if the program leaks out of the writer's or the consumer's office, anyone can copy the program for the price of a blank diskette. In such a world, very little software will be written. The reasons are twofold:

1. Few software applications generate enough benefits to justify financing all the fixed costs of software creation.
2. Even for those applications that do exist, the potential financier may adopt a wait-and-see attitude; he will wait for someone else to buy the program so that he can copy it.

The economic jargon for this behavior is "free riding" if he gets the program for free. To the extent that he shoulders some of the cost but shoulders less than the average cost, he is "easy riding." It is in the writer's best interest to devise mechanisms to exclude free and easy rides.

Free rides and easy rides should upset us all. Not only are they unethical, but they also translate into inefficiency and the misallocation of resources. With respect to software, many potential programs could generate value greater than the costs of writing the programs. Or to put it in the language of economics: each potential consumer could increase his revenues by more than the average fixed costs of software creation. Nevertheless, because of free and easy riding, little is bought and so little is written. National output will be lower than if there were no free and easy rides, and we will all be worse off.

In a world without property rights over software, free riding will be rampant and the writer will not be able to reap the fruits of his labor. Will he still produce software anyway? Well, what is software? Software is a product of the writer's education, capital, and time. However, his education, capital, and time are not locked into writing software. All can be allocated to products or services for which he can be paid. For example, the writer could have trained as an architect; no one except a paying customer would receive a blueprint. But as a writer in a world without property rights over software, anyone can enjoy the fruits of his labor.

Governments recognized the potential for such market failure years ago and quickly extended copyrights to software.[5] However, rights de jure are not enough; money also has to be spent to police the property—a very significant transaction cost. So software companies routinely advertise that pirated software is theft and will be prosecuted. However, because policing costs are so high, software is still pirated. Recognizing this, software companies have also invested in the development of ingenious exclusionary mechanisms to prevent more than one copy of a master diskette. But again, property rights are never fully delineated. Over time, hackers have invented ways around the exclusionary mechanisms!

To complete the analogy, consider the landowner and the land. Landowners who own habitats and leave them undisturbed are generating a benefit for industrial users of that information just as a software writer is generating a benefit to users of his program. The industrial users could buy the habitats where the genetic information resides. But there will be precious few users who will derive enough value over any one commercial application of a GCF to make it profitable to acquire the whole habitat—just as there were precious few software users who derive enough value from the application of a program to finance the creation of software. And just as we expect little software written in a world without property rights over software, we should expect little conservation in a world without property rights over genetic information.

When one considers transaction costs, the analogy is probably a better argument for property rights over genetic information than it is for property rights over software! Consider, for example, the ease of copying information and free riding. Although it is hard to put a fence around information, be it on floppy disks or in a habitat, the difficulty is not equivalent. It is much easier to police entry into classified software than entry into habitats. Software can be locked into computers, and access can be limited. Witness the CIA, the former KGB, or any secret police. All have spent large sums of money to erect physical barriers to exclude the entry of hackers. The same cannot be done with genetic information. Erecting fences around thousands of hectares of remote jungle not only is not practical but could even exacerbate the mass-extinction crisis through isolation and inbreeding.

The analogy between copyrights over software and intellectual property rights over genetic information demonstrates the importance of incentives. Presently, there is no incentive on the part of landowners to preserve genetic information. Under international law, genetic information is considered "the common heritage of mankind," even though the land is not. Landowners, like software writers, do have a claim to the more excludable goods and services that can be wrought from their capital. Just as the writer could have become an architect, the landowner could deploy his land for a number of things: hydroelectricity, cultivation at all scales, timber, and mineral extraction—to name a few. Economists call such forgone benefits "opportunity costs." Because of these positive opportunity costs combined with the prohibitive transaction costs of capturing the benefits from GCFs, one can only conclude that genetic information will continue to be expunged. But if the transaction costs can be lowered to where the benefits from the GCFs are greater than those from alternative land uses, it will then become in the selfish interests of landowners to conserve genetic information. The rest of this book will examine the most significant transaction costs and elaborate how government can reduce these costs and allow selfish interests to drive conservation.

Although the logic of the policy is somewhat obvious and can even be illustrated by an analogy, the solution is by no means simple. The complexities of the institutions involved—intellectual-property law, landownership patterns, contractual forms, international finance, tropical taxonomy—require a very careful and tempered analysis. In other words, the solution is complex. The chapters to follow will tackle these complexities and attempt to break them down one by one into a practical solution.

5

Illustrating the Problem
and the Property-Rights Solution

Theory is abstract, and abstractions, by their nature, are difficult to comprehend. Many astute readers will not have the patience to keep track of the definitions and acronyms, plod through the sequence of steps, and deduce the solution. For many, privatization may still seem too abstract to work. As an economist engaged in the enterprise of rhetoric, my job is to persuade multiple audiences that the solution is workable. In other words, I must persuade the readership that privatization does not teeter on abstractions; privatization is both simple and practical. Indeed, it is so simple that the essence of the argument can even be communicated through illustration.

There are many reasons for illustrating both the problem and the property-rights solution. The first is that good illustrations are universally recognizable, while words are not. Many people living in the remote tropics can barely read or write. Inasmuch as these same people are the most affected by any conservation policy, they should know the basic mechanism of the policy and have some say in its design and implementation. This is not just some lofty democratic principle that I espouse (although I do); it is a deductive consequence of the property-rights approach. A policy that poor people see as one in their own best interest will entail the lowest transaction costs in policing, whereas a policy that poor people do not see as in their interest will be frustrated whenever the poor enjoy some de facto property right over the resource allocated by the policy. To discover the most efficient policy, technocrats in the North would do well to speak and listen to the poor of the South. Ideally, an array of policies should be floated among the poor and the illiterate.

There will be holes in any policy floated. The holes that affect the rich and articulate will be spotted immediately and communicated—loud and clear—by the rich and articulate. These holes will be plugged. But the same cannot be said of holes that affect the poor and uneducated. Poor people usually have no impact on policy making. Nevertheless, once a policy is chosen and implemented, the poor will react and circumvent any part of the policy not in their interest. In the case of conservation, they will react through squatting, wildcat prospecting, and poaching. Therefore, to ignore the poor seems quite foolish. It makes much more sense to get the feedback of representatives from all affected parties—including the poor and uneducated. Informed trade-offs can then be made.

How can a conservation policy enjoy public comment from all parties concerned? The answer is not so complex: we find an appropriate medium. The usual media—television, radio, newsprint—are fine for the higher rungs on the socioeconomic ladder. Fortunately for the policy, these rungs have the most power and can be persuaded at low cost through these media. It is also fortunate, for humanitarian reasons, that these rungs do not have complete power. Poor people will have some power to frustrate the policy as long as the policing costs to prevent squatting, and other land uses, are higher than the value of conservation forgone. Therefore, the policy must be communicated to the poor, and the poor must be persuaded of its value to them.

The thesis of this book is that the creation of property rights over genetic information can make habitat preservation compete with alternative land uses. Conservation will be chosen whenever the royalties paid are greater than the benefits of other land uses. In such cases, the managers of habitats will become the residual claimants to the benefits. If the poor are made these residual claimants, then they will have an incentive to preserve.

To communicate the profitability of the policy requires ingenuity inasmuch as the usual media are not appropriate. Indeed, the rural poor often do not have access to electricity, much less a television or a radio. And as for newsprint, well, the poor often cannot read. Nevertheless, the poor may have access to one who can read and explain a diagram printed in the newspaper or distributed by pamphlet. A policy that is diagrammed and well explained will probably be remembered; the pictographs of the diagram can jog the memory.[1]

Inviting the poor and uneducated into the decision-making process is a radical idea. Economists in institutions like the World Bank will no doubt ridicule it. They should not, especially in light of their own poor performance.[2] A very strong causation exists between the bank and the current mass-extinction crisis. The causal link is the financing of highways, dams, timber operations, and agriculture in virgin habitats. To add insult to injury, many of these loans cannot be paid back even at the artificially low interest rates charged by the bank! In other words, the bank's economists grossly miscalculated the profitability of the projects and squandered taxpayer dollars. And taxpayers and the biota were not the bank's only casualties. To recoup for bad loans, the bank has imposed austerity programs that are effectively keeping one-fifth of the world population in clinical starvation. Neither "food first" nor "all things great and small" has ever been the motto of the bank.

This criticism of the bank may seem tangential to illustrating the problem and the property-rights solution. It is not. Many of my colleagues will argue: persuade the bank and forget about communicating to the poor and uneducated. I am arguing: persuade the poor and uneducated and forget about the bank (at least for now). I feel it would be folly to pin one's hope on persuading the bank. The reason goes back to the overarching theme of this book—selfish interests.

Privatization as a conservation policy does not correspond to the incentive structure of the bank.[3] In fact, it goes against two driving forces within the bank: (1) powerful lobbies vying for large-scale capital projects and (2) the bank's bureaucrats, who are rewarded for approving overblown projects and belittled for even considering small-scale enterprises. One does not have to be a cynic to suspect that privatization is a piece of rhetoric embraced by the bank only when it corresponds to the selfish interests of its bureaucrats and lobbies. I hope that I am wrong. However, I believe that I am not. Therefore, the omission of the World Bank in the schema to follow is no oversight; it is deliberate.

Figure 5.1 is an attempt at illustration. The left-hand side illustrates the economics of deforestation and mass extinction. On the top, just left of center, is the rain forest, represented by treetops and clouds. Below the forest are those who control the forest and the fate of the species in it. Some of these individuals are the strongmen of what amount to feudal empires; others are squatters eking out a bare subsistence, also in a feudal way. For lack of a better term, all these individuals are called "landowners."

Landowners can sell the forest for various commercial purposes: small- to large-scale agriculture, mineral exploration, amd hydroelectric dams. When the forest is bulldozed, money flows to the landowners in the form of rents, to governments in the form of taxes, and to corporate boards in the form of profits. It is in the selfish interest of each party to deforest.

Below the landowners is the government (the domed building). Landowners influence government, and governments influence landowners.

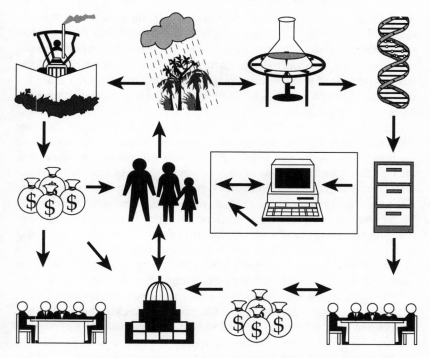

FIGURE 5.1. The economics of deforestation and preservation.

Hence the double-headed arrow. But the influence is never absolute. Even in the First World, governments cannot adequately enforce endangered-species laws. And often when the laws are enforced, landowners challenge the restrictions as a "taking" of property without "just compensation."[4] In the Third World, governments have far less control. In some countries, governments have virtually no control. As mentioned in the previous chapter, the map of Africa is dotted with "paper parks," and the Brazilian Amazon, like all new frontiers, lives by the law of the gun. The unhappy reality is that landowners, acting in their own self-interest, will deforest virgin habitats and exterminate species irrespective of laws protecting habitats and/or species.[5]

Now consider the right hand side of the diagram. Start again with the clouds and treetops, but ignore for now the boxed computer terminal in the center. Moving right from the rain forest along the perimeter of the diagram, we see that industries can assay or extract genetic information from specimens collected in the rain forest. Production processes that use the information are patented and the revenues generated remit to the corporate boards in the form of profits and to governments in the form of taxes. In this clockwise circle around the right-hand side of the figure, the landowners are left out of any direct money flow, even though they have control over the existence of that information!

Let us consider the selfish motive of each agent in the figure. Because the landowners receive nothing for preserving genetic information, the benefits to the landowners in the left-hand side will always be greater than the benefits from the right-hand side. In other words, the landowner's have no incentive to preserve. Deforestation and extinction are also in the selfish interest of some corporate boards (e.g., mining and large-scale agriculture). Those boards are to the left. However, other corporate boards (e.g., in pharmaceuticals and agribusiness) will want to preserve the rain forest as a bank of genetic information. Those boards are to the right.

In both the left- and right-hand sides, government receives money from taxes paid. If good government acts by the simple maxim, "produce the greatest good for the greatest number," government policy should favor the value of the land use that is greater.[6] From the government's viewpoint, the issue to preserve or to deforest reduces to an empirical question: Which value is greater? Based on some simplifying assumptions, one can calculate the potential value of the rain forest for the pharmaceutical industry at $300 per hectare a year and the average alternative land uses at $170 per hectare a year.[7] With $300 > $170, the value of the rain forest *as a bank of genetic information* is greater than its value in alternative land uses. The trick for government is to devise a mechanism that produces the greatest good for the greatest number—that is, the $300+ per hectare a year derivable from preservation.

At this point, an economic explanation must accompany the figure. Central to property-rights analysis is the efficiency criterion: *those who control an asset should also be the ones who derive the benefits from that asset.* The people below the rain forest control the rain forest. To make them the beneficiaries requires a mechanism to exclude the corporate boards to the right from benefiting whenever they have not yet paid the costs associated with the benefit. In the previous chapter, the exclusionary mechanism for software was copyright protection and its enforcement. An analogy was made between software and genetic information. Given the closeness of the analogy, it is not surprising that the exclusionary mechanism for genetic infor-

Privatization Will Be Profitable

In *A Wealth of Wild Species*, Norman Myers estimates that (1) there are 50 times more drug-offering species than being used today and (2) species-related drugs yielded $40 billion per year.* Myers's estimates portend a potential of $2 trillion per year given present technologies. However, biotechnologies are revolutionizing industries and are intensively using genetic information. What this means is that the $2 trillion per year may be a gross underestimation. Of this $2 trillion, what percentage can be contributed only by a specific piece of genetic information? This is the thorny issue of substitutability in microeconomic theory: for cough medicines, substitutability may be high; for many chemotherapeutics, it may be close to zero. Inasmuch as genes are information, one can justify the assumption that the percentage contribution is roughly the same as that which artists receive for the information they create: 15 percent (However, Madonna Louise Ciccone signed a contract in 1992 with Time-Warner for 20 percent, and Michael Jackson got 22 percent from Sony Entertainment in 1991.) On $2 trillion per year, 15 percent renders a royalty value of $300 billion per year to landowners of virgin habitats. With 1 billion virgin hectares extant worldwide, royalties would average $300 per hectare a year.

What are the opportunity costs of preservation? This is another thorny microeconomic issue. However, there is a simple way to approximate an answer. The 1990 OECD estimates placed the total Third World debt at $1,322 billion. The total debt service was $170 billion, and most countries in the South could not meet their debt service. When divided by the billion hectares of virgin forest, the debt service translates to a mere $170 per hectare a year. Privatizing genetic information could not only cancel the debt, but also provide a profit margin of $130 per hectare a year.

*Norman Myers, *A Wealth of Wild Species: Storehouse for Human Welfare* (Boulder, Colo.: Westview Press, 1983).

mation should also be a law and its enforcement. Legislation should be drafted in both the North and the South to create intellectual property rights over natural information.

The essence of the law would be simple: the owners of any new product that utilizes genetic information must compensate the owners of the habitats where that genetic information is distributed. In the case of pharmaceuticals, this means that patent holders will pay royalties to the landowners who provide the service of preserving genetic information. Ultimately, these royalties will be embedded in the cost structure of the product and be borne

---

Equal Protection of Artificial and Natural Information

To protect natural information, preservers of natural information will enjoy rights equivalent to those that govern patents, copyrights, semiconductors, trademarks, and trade secrets. Any good or service that cannot be produced without the utilization of specific natural information will remit to the preservers of that information 15 percent of the value added to sales, except for trademarks where the royalty will be 1 percent. Each preserver will enjoy a share of royalties proportional to the share of the total natural information he controls. For a patent, these rights will extend 17 years from the date of patent. For copyright, these rights will extend for the life of the author plus 50 years; preservers will enjoy a share of royalties for that part of the work which utilizes the natural information proportional to the size of the work. For semiconductors, the rights shall extend 10 years from the date of registration of the semiconductor. For trademarks, the preservers will receive a royalty as long as the company utilizes the natural information in their trademark. For trade secrets, the commercial use of natural information will remit royalties to the first known utilizers. Any natural information already commercialized prior to the legislation will lie in the public domain.

---

by the consumer. Consumers should not complain. Recall the baseline ethic agreed to in Chapter 1: *those who benefit pay the costs associated with that benefit.*

What cannot be done with the stroke of a pen is the actual biological inventory. Exclusion works only if it reveals true ownership. Because habitats do not correspond to landownership patterns, royalties will have to be shared among landowners. A careful inventory is a necessary condition for determining who gets what.

The inventory is illustrated in the diagram Figure 5.1 as a computer terminal. The input of information into the inventory is illustrated with arrows. Landowners register the genetic information on their property with the inventory, and corporations reveal the genetic information used in their products to the inventory. The inventory makes the match to identify claimants. When a large sum of royalties accrue, ecologists go out to the field and measure the extent of each landowner's population of the organisms containing that genetic information. Royalties are then paid proportional to the number of organisms in the taxon in which the function is genetically coded. The particulars of the biological inventory and its financing are developed in Chapters 7 and 9, respectively.

There are huge practical problems. Most of the species in the rain forest have not yet been identified, much less inventoried. One tropical biologist estimates that it would take a decade to classify the species on 1 hectare of Costa Rican rain forest![8] Huge costs will be entailed in the creation of a biological inventory. To reduce these costs, landowners can negotiate with biologists a share in royalties. If landowners do not want to assume these risks, then they can sell their rights to corporations or "genesteaders" who will.

6

Genesteaders

Let us suppose that governments around the world legislate ownership over genetic information. Under the proposed law, those who conserve genetic information would become the residual claimants to the royalties. Because conservation does not just mean letting the habitats be, the policy will require managers. Several issues quickly emerge. The first is one of desirability. It boils down to this: How many people really want the job of genetic-information management?

Economists will step in and claim this as their turf. They will see the question as of one of "elasticity," where elasticity is defined as the percentage change in quantity divided by a percentage change in price. In this case, what percentage of genetic-information managers will offer their services for what percentage of expected royalties? Such a calculation would be a mechanical exercise if the data existed; but the data do not exist. What's an economist to do? The answer: assume. The economist could assume that the labor response to royalties will be similar to the labor response to, say, the gold rush in the Amazon for which the data do exist.[1] To the extent that the assumptions deviate from reality (i.e., the gold prospector is psychologically different from the genetic-information manager), the elasticities will be bogus. Nevertheless, the economist will be paid, and he will make his usual disclaimer about test assumptions and *ceteris paribus*.

Although an honest answer to the quantitative question How many people really want the job of genetic-information management? is impossible, a qualitative answer may well be within the reach of evolutionary theory. Evolutionists argue that modern activities are pleasurable to the extent that they simulate the activities of human's evolutionary past. The explanation is that individuals who enjoyed those activities would perform them better and in performing them better, they would have transformed more of their environment into themselves and their offspring. In other words, nature selected genes that code for pleasure in activities that enhance survival. Inasmuch as man evolved as a hunter and gatherer, modern activities that simulate hunting and gathering should be pleasurable.[2]

Such evolutionary reasoning can shed some light on the issue of how desirable the job of genetic-information management will be. Two prominent evolutionists, E. O. Wilson of Harvard University and William Hamilton of Oxford University, have independently hypothesized that natural selection may have equipped humans with a genetic biophilia[3]—a love of life. From the evolutionary perspective, Wilson and Hamilton speculate that hominid abilities to assess biodiversity were once highly adaptive. In other words, genes for biophilia may have enhanced the survival of individuals possessing those genes. Evidence that biophilia enhances survival appears in almost all ethnographies of modern hunting and gathering groups. Members of such groups are taught at a young age which species are useful for food, tools, shelter, clothing, and medicine. This knowledge is culturally coded and highly fragile. That genes for biophilia would have enhanced survival in human evolution is buttressed by primatological evidence. Jane Goodall has documented an appreciation of biodiversity in the chimps of Gombe Stream.

If Wilson and Hamilton are indeed correct in their speculation, then genes for biophilia should trigger pleasure in conservation activities. The impact of these behavioral genes will be felt in the willingness of labor to engage in genetic-information management. To the extent such management may satisfy deep human wants, an eager supply of managers will be forthcoming at relatively low compensation. Nevertheless, to quantify this qualitative response with the precision of a number would be misleading, if not dishonest. A meaningful number is simply beyond our lens of resolution.

This chapter will explore the organization of genetic-information management, given a willing labor supply. Fundamental to the discussion is efficiency. By the efficiency criterion of property-rights analysis, *those who control an asset should be the ones who derive the benefits from that asset;* the Southerners *who control the habitats should be the ones who derive the benefits from the habitats.* Since property rights over genetic information are tied to habitats and habitats are tied to the property rights over land, justice must first determine who holds title to the land. When justice determines that indigenous peoples hold title to the land, then the new property rights over genetic information should reside with the indigenous people. When

If GCFs, Why Not CCFs?

The pharmaceutical giant Eli Lilly produces two chemotherapeutic drugs derived from a rosy periwinkle endemic to Madagascar. Eli Lilly did not discover the rosy periwinkle accidentally; researchers surveyed medicines used by shamans to treat topical rashes—an indication of chemotherapeutic activity.

The shamans of preliterate cultures are walking *Physician's Desk References* (*PDRs*). However, their knowledge of medicinal plants differs in a crucial way from the indications/contraindications of the *PDR*. The *PDR* can be photocopied; the shamans' knowledge can only be transmitted orally. Whereas the chemotherapeutic value of the rosy periwinkle was a genetically coded function, the knowledge that the rosy periwinkle has medicinal value is a culturally coded function (CCF).

In many respects, CCFs are more threatened than are genetically coded functions. Tribal youths are abandoning their cultures and communities for the city-slum life. Often there is no one for the shaman to teach. How to preserve these CCFs?

A debriefing from shaman to botanist is the cost-minimization solution.* However, there is a problem. Why should a shaman share his *PDR* with a Westerner? Surely not altruism: the horrific treatment of native peoples is a history still in the making. Is there a solution to retrieve CCFs to the benefit of both Westerners and preliterate tribes?

The answer may again lie in property-rights analysis and privatization. The knowledge of which species exhibit medicinal properties often resides in one person or a small group of people within the tribe. If the tribe incorporates, then the identification of medicinal plants can be viewed as intellectual property, a type of trade secret. And if a tribal healer is induced to divulge the tribe's secret to one of the pharmaceutical giants, then trade-secret law would come into play. The healer would have committed a breach of contract, and/or the Western corporation that has acquired the information would have committed a crime like espionage or bribery. When it comes to assessing damages under U.S. trade-secret law, the courts will take account of the commercial value of the trade secret. Once shaman knowledge is legally recognized as intellectual property, tribes can sell their CCFs for royalties—royalties that could finance ethnographies and/or induce tribal youth to remain tribal.

*William K. Stevens, "Shamans and Scientists," *New York Times*, 28 January 1992, B1, B9; Michael Balick, "Ethnobotany and the Identification of Therapeutic Agents from the Rainforst," *Ciba Foundation Symposim* 154 (1990): 22–39.

justice determines that urban investors hold title, then these investors should be able to sell the newly created rights if they so wish. When justice determines that government holds title, then the titles themselves should be disbursed according to the most efficient sized land parcel for genetic-information management.

Inasmuch as justice is administered by government, it will probably be government that determines that it has title to the land. This may seem cynical, but I believe it is realistic. Once government secures titles to virgin habitats, the titles should be disbursed to entrepreneurs who will serve as genetic-information managers. I call this new breed of entrepreneurs "genesteaders" in conformance with the nineteenth-century U.S. institution known as homesteading. But the term "genestead" in no way pays homage to the institu-

---

Monkey Know-How

The remarkable Jane Goodall has documented that ill chimpanzees pick leaves of plants known to possess therapeutic effects.* This function is not genetically coded. A chimp bred in captivity and reintroduced into Gombe Stream would have no more idea as to which leaves to chew than would you or I. The pharmacognosy of the chimps is a culturally coded function (CCF).

Like almost everywhere else, Gombe Stream has been encroached by human settlement. Today, the CCFs of chimps are very much threatened. Is there a solution to retrieve these CCFs to the benefit of both chimps and humans? The answer is again privatization, but with a twist. Unlike the shamans of Madagascar, the chimps of Gombe Stream can never enjoy human rights, much less the right to litigate trade-secret infractions. Nor can the chimps easily transmit the highly valuable information of medicinal plants. The CCFs of chimps have been ascertained only by the painstaking primatological research of Goodall and others—research that has spanned decades. Who should own these CCFs? By the efficiency criterion of property-rights analysis, *those who control an asset should be the ones who derive the benefits from that asset*; the legal title should reside in the researchers of Gombe Stream.

At this moment, researchers at Gombe Stream are investigating 27 other species rarely eaten by chimps. The economic advice to Goodall is that she keep her findings secret until privatization eventuates. The royalties can then be used to buy contiguous lands and study chimps.

*Jane Goodall, "The Chimpanzee's Medicine Chest," *New Scientist*, 4 August 1990, 26–28; James Grisanzio, "The Monkey's Medicine Closet," *Technology Review*, August–September 1993, 13–14.

tion of homesteading. Before delving into the organization of the genestead and the remuneration of genesteaders, let me first explain the rhetoric. The neologism "genestead" forces me into an apology. Just as "privatization" is saddled with unfortunate connotations, so too is the word "genestead." The term "genestead" is analogous to the word "homestead," and I coin the word with some reluctance. Homesteading was a federal program to privatize vast expanses of territory. The rationale for the enabling Homestead Acts of the mid-nineteenth century can be best understood in the context of Manifest Destiny—the doctrine of a United States of America stretching from sea to sea. The stretch was inspired by God. The phrase became the rallying cry for the United States to launch aggression against its neighbors and to justify subsequent land grabs. To secure new borders, the United States achieved destiny through enfranchising pioneer families to settle and "improve" the land. Improvement was a dual concept: it meant "self-defense" from Amerindians (today we call it genocide) and clearing the bush (today we call it ecocide).[4]

Given the sad eco-history of American homesteading, genesteading may seem a poor choice of words. However, the conceptual basis of genesteading is virtually identical to that of homesteading: the government cedes title to land if certain conditions are met by private individuals. In the case of genesteading, the "improvements" are virtually the opposite to those of homesteading. Genesteaders will gain title to royalties by performing habitat management and classification tasks. These improvements will protect native organisms.

To envision a genestead, one must envision a tract of land—its location, size, and tenure—all with the ultimate end of conservation in mind. This is analogous to the concept of the homestead. To envision a homestead, one envisions a tract of land—its location, size, and tenure—all with the ultimate end of land use in mind. Because the two ends are so inverted, the very way one looks at the location of the genestead must also be inverted. With the homestead, one looks out from dense human inhabitation as it thins and approaches wilderness. With the genestead, one looks from the heart of the wilderness and sees how it thins and approaches human inhabitation. From the genestead perspective, the habitats most threatened by deforestation are those at the ecological edge. In other words, the frontier is the land at the edge of virgin land. Although the frontiers of the homestead and the genestead are the same edge, the way of looking at the location is so different that it will imply different optimal sizes and tenures.

The Brazilian rain forest is a good setting to illustrate the location, size, and tenure of a genestead. The Amazon is a network of rivers traversed by an occasional highway penetrating the bush. The rivers and roads have become conduits for deforestation and extinction.[5] Although the causes for deforestation are multiple, some are as simple as hope and human survival. Peasants escape impoverished northeastern Brazil and penetrate the interior on public buses up the Trans-Amazonic highway. As homesteaders, the peasants "improve" the land by virtually the same means and with virtually the same ends as their nineteenth-century American counterparts. Privatization as a conservation policy attempts to take the same labor that would homestead a tract of land and provide incentives to genestead it.

The genestead is illustrated in Figure 6.1, which is generic and could depict any road or river system in the tropical Third World. To make it real, I have chosen the clinical name BR 364, an actual road in Brazil's westernmost state, Rondônia. Imagine that the thick black line down the middle is the road. To either side is virgin habitat. The dotted lines are the proposed boundaries of government property ceded to the genesteaders. It is assumed that beyond the outermost boundary is government property.

The genesteader's tasks are (1) to protect a strip of land along the road from peasant squatting, (2) to engage in simple habitat-management techniques, and (3) to assist in the classification of genetic information (the details

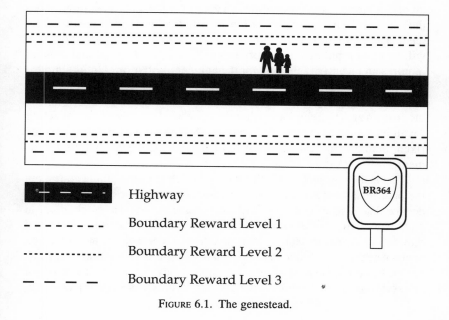

Highway

Boundary Reward Level 1

Boundary Reward Level 2

Boundary Reward Level 3

FIGURE 6.1. The genestead.

of which will be elaborated in Chapter 7). In return for performing these duties, the genesteaders will earn royalties on any genetic information on their genestead that renders commercial value. The royalties will be in proportion to the copies of that piece of genetic information within the boundaries of the genestead vis-à-vis the population of that same piece of genetic information on all other properties (in accordance to the legislation suggested in Chapter 5). The particulars of shared royalties and the tenure of claims will be deferred to Chapter 9. Of interest here are the boundaries. Why are there multiple boundaries emanating from the road?

The three boundaries are merely suggestive of a continuous reward structure that cedes more title for varying degrees of job performance. There could be a dozen or more boundaries emanating from the road according to the number of genestead activities that are cost effective to monitor. The boundaries reward the genesteader for his level of performance. The reward is through the expected claim to royalties; a bigger genestead means more organisms on it and a greater likelihood of future royalties should the genetic information of an organism yield commercial value. Due to the limitation of graphical presentation, only three boundaries are depicted.

Somewhat counterintuitively, the greatest reward should be given for an activity that has nothing to do with protection against squatters, habitat management, or classification. That activity is family planning. Given the exponential nature of population growth, family size will greatly affect the conservation of genetic information in the long run. This means there will be an optimal number of offspring. To achieve that optimal number requires an incentive schedule for the number of offspring. In its simplest form, the schedule moves the boundary of the genestead according to the number of offspring. For the optimally sized family, the boundary would move outward, thereby increasing the genesteader's likelihood of future royalties. Once all the land is "genesteaded" and the boundaries become fixed, then the percentage of royalty becomes the mechanism of adjustment.

What is the optimally sized family? Although this is a complex question, the precise answer is not beyond the scope of property-rights analysis. To deduce that answer, one starts with an extreme suggestion: no children. In fact, many hard-core environmentalists claim that zero children is indeed optimal. They reason: let the planet breathe and repair before embarking on another generation. Property-rights analysis suggests something very different. From the vantage of property-rights analysis, the genestead is just an asset from which benefits flow. The flow is only as secure as it is worthwhile to defend the asset from encroachment. Defense costs money. It is quite plausible that offspring would actually lower the transaction costs of policing the genestead against encroachment.

To clarify this reasoning, imagine two reward structures corresponding to the views of hard-core environmentalists and those of property-rights analysts. The first incentive structure gives the greatest reward to genesteaders who have no children. Those genesteaders exercising the first option would find their genestead and themselves vulnerable as they age; squatters, poachers, and alien species would invade their property. In other words, the policing costs of protecting native organisms from squatters and other intruders would rise. At the same time, the aging genesteaders would most likely be less responsive to incentives to preserve, as a consequence of their childless affluence. Or to put it in the language of economics, their labor would become more inelastic.

In contrast, the genesteaders opting for one child would have an incentive to manage the genestead well so that their only child would inherit an asset that continues to generate a royalty stream. Their labor would remain elastic. The only child could also manage and police the genestead as the parents age. If one is better than none, is two better than one? How about three? four? More than one child incurs the transaction cost of negotiation of inheritances as well as disruption of the optimal size of the genestead. A one-child rule seems optimal. Nevertheless, the implementation of such a rule would be complex. How do we implement a structure that rewards genesteaders more for having one child than for having none (or two), two rather than three, three rather than four? With great care! A poorly designed reward structure, dreamed up in the North and insensitive to the cultures of the South, may ultimately defeat the whole policy of privatization. For example, incentives for one-child families may create incentives for female infanticide and/or prenatal screening of sex as a basis for abortion.[6]

To the question of infanticide and abortion, the cynic may say: So what? With nearly 1 billion people on the planet already starving or near starvation, why should one care about infanticide and/or abortion?[7] The answer to such a cynical question is an equally cynical response: politics. Any part of a policy that unwittingly rewards infanticide and/or abortion will be fiercely opposed by the Catholic Church and by Protestant fundamentalist denominations. Religious groups may come down hard on the whole policy because of the implication of just one part. And these groups will not be the only obstacle: any part of a policy that constrains reproductive freedom may also be opposed by feminists.

Shoulder to Shoulder

In the long run, privatization cannot be divorced from family planning. To illustrate this, consider some simple calculations. Imagine (1) a 160-hectare genestead, (2) a family lineage in which each couple has three children in each generation, and (3) 1 square meter of land, which can hold three standing adults. One hectare is $10{,}000m^2$ or $0.01km^2$, so 160 hectares is $1{,}600{,}000m^2$ or $1.6km^2$. A human generation is 25 years. Can one extra child per generation continue ad infinitum? No! There are only $1{,}600{,}000m^2$ on the genestead and each meter can hold only three people or, conversely, each person occupies $0.333m^2$. Some future generation n will find the genestead shoulder to shoulder with people.

When does n obtain? To simplify the calculation, think of procreation on a per person basis. The two parents who had three children, had, on a per person basis, 1.5 children—half a child more than replacement. For how many generations (n) can this go on until the genestead is shoulder to shoulder with people? One can calculate this as 1.5^n persons $\times 0.333m^2$/person $= 1{,}600{,}000m^2$. After some algebraic manipulation, the answer is $n = 38$, or only some 950 years!

To reduce the probability of perverse responses to family-planning incentives, some simple guidelines should be followed. The first is that the reward structure for the single male must be neutral as to his civil status. That is to say, the male should receive the same size genestead whether he marries or does not marry. However, if he marries, then his wife should receive a contiguous lot. Since reproductive decisions ultimately reside in women, by the efficiency criterion of property-rights analysis (*those who control an asset should be the ones who derive the benefits from that asset*), women should be the ones who expand the genestead according to the number of offspring. Say both the man and woman receive 160-hectare plots configured as strips along the highway. If, over her reproductive life, she has only one child, then her strip trebles to 480 hectares; if two children, then it doubles to 320 hectares; if three children, then it is just the 160 hectares; if four children, then it halves to 80 hectares; and so on. With a high reward for one child and economies of scale in performing genestead activities on the contiguous lots, men will have a strong incentive to see their wives enfranchised and their offspring recognized.

After family planning, there are two basic activities of genesteading that require a great deal of elaboration: habitat management and classification. The skeptic may now query: How can marginally literate people perform habitat management and/or classification—activities that require formal study and even advanced degrees?[8] The answer: they cannot. Here, again, language becomes a sore point. The habitat management is not of a sophisticated sort. Neither is the classification. The reader should dispense with the connotations those terms have in the North.

Habitat management will mean following simple instructions and learning routines. The instructions and routines can be established by scientific

---

Constraints on Reproductive Freedom: Environmentalism versus Feminism?

Most environmentalists would agree that the only way to save the planet is to limit human population size, control affluence, and adopt appropriate technologies. In their book *The Population Explosion*, Paul Ehrlich and Anne Ehrlich conceptualize the relationship between the degradation of the environment and its causes as $I = PAT$, where the I is environmental impact, P is population, A is affluence, and T is technology.* Of the three variables, P is the most worrisome; large gains in controlling A and T can be wiped out by an exploding P. To limit population size means to limit family size. This usually happens voluntarily as the standard of living rises. However, a rise in living standards takes place in the long run, while environmental degradation takes place in the short run. The logical conclusion is that family size must be limited involuntarily. This means constraints placed on reproductive freedom.

How will the political spectrum react to such constraints? The position of the Catholic Church needs no elaboration. However, the position of feminists does. Feminists argue pro-choice on the basis of reproductive freedom. However, reproductive freedom is as much the right to reproduce once, twice, or thrice as it is not to reproduce. If one allows environmentalists to constrain twice or thrice, why cannot anti-abortion advocates constrain zero? Both positions are based on the rejection of reproductive freedom. I had the good fortune to put this question to one of the founders of the modern American feminist movement, Betty Friedan, during a public lecture. Her response: this is a "false problem." When pressed, Friedan dismissed the question as a ruse for some other agenda.

*Paul Ehrlich and Anne Ehrlich, *The Population Explosion* (New York: Simon and Schuster, 1990), p. 58.

Labor-Intensive Genesteading

Newly arrived species can become pests. How to get rid of them? One solution is the introduction of a predator species. Unfortunately, this solution is almost always the wrong one. For example, the early-nineteenth-century colonizers of Hawaii introduced rats (unwittingly, of course) through ship cargo. Rats became a pest and posed a human health hazard. So the mongoose was introduced. Then the mongoose drove endemic birds species to extinction through the eating of their eggs. The mongoose became a pest, and so the cat was introduced. Then the cat population exploded, and cats became a pest . . . unfortunately, there are no short-cuts. A species introduced by man must be eliminated by man; elimination is a labor-intensive activity. To take a short-cut and introduce another predator is to assume knowledge about where that predator lies in the ecosystem, knowledge that ecologists just do not have. Invariably it backfires.

Another labor-intensive activity is pH restoration of habitat. An example is the habitat of the uncharismatic mini-carnivores—the batrachians. The batrachians are the frogs, toads, and salamanders. Many batrachians produce toxins that hold promise for medicine. They are all marching to extinction even in habitats far removed from any point source of pollution.* Their decline is attributed to the overall acidification of the environment. It is unrealistic to expect acid rain to abate sufficiently to save many of the batrachians—intervention must happen now to restore the proper pH. But what incentive does a landowner have to "lime" a pond and carefully measure its acidity or alkalinity? Presently none. Privatization would give the landowner the incentive of royalties.

*Richard Griffith and Trevor Beebee, "Decline and Fall of the Amphibians," *New Scientist*, 27 June 1992, 25–27.

---

experts from the North in concert with education experts from the South. Some of these tasks will be simple yet labor intensive, such as removing nonindigenous species or restoring soil pH.[9]

Classification will also mean following simple instructions and learning routines. The importance of classification is crucial to privatization as a conservation policy. Only through an accurate classification can one determine who will get what, should the genetic information of an organism on the genestead have commercial value. To the genesteader, accuracy in taxonomy can mean either boom or a modest livelihood. The reason for accuracy is boom: finding a rosy periwinkle (the plant which rendered vincristine and vinblastine) on one's genestead and one's genestead alone makes one a multimillionaire. But usually, the distribution of genetic information will not correspond to existing landownership patterns; royalties will have to be shared. Who shares what will have to be determined by a gargantuan database of land titles and habitats.

7

The Rationale, Design, and Implementation of the Gargantuan Database

Delineation of land is the creation of titles. Most of the dry surface of the Earth is delineated to some degree. The exact degree is predictable by a simple criterion of property-rights analysis: if the benefits are greater than the costs, then the land is delineated and a boundary emerges.

The effect of delineation is a reduction in negotiation costs. Uncertain boundaries dampen the incentives to invest in land potentially under dispute. In such cases, ownership over the value added to the land may some day have to be negotiated. For example, no homeowner should enclose his yard without first surveying the property. On a grander scale, no government should permit land improvements near a border if the border itself is in dispute.[1]

One way to lower the negotiation costs of ownership is to use simple signals to delineate the land. The rivers of the United States are a good example. They separate the United States from other nations, states from states, cities from cities, and even boroughs within cities from other boroughs. For example, the Rio Grande separates the United States from Mexico; the Mississippi River separates all the states to its east from all the states to its west; the Delaware River separates Philadelphia, Pennsylvania, from Camden, New Jersey; and the East River separates the borough of Manhattan from the boroughs of Queens and Brooklyn in New York City.[2]

Habitats also have their boundaries. Sometimes the boundaries are visible barriers like oceans, mountains, and deserts. But usually the boundaries are less visible. Things like pH, temperature, and aridity can delineate the boundary of a habitat. Over time, the differences of individuals within a species with respect to their tolerance levels of pH and other factors will allow some to radiate beyond these boundaries and evolve into races and even new species.

Because the boundaries of landownership and those of habitats are usually not coincident, species will inhabit many parcels of privately held land. Consider the Rio Grande. Although the river separates Mexican and U.S. jurisdictions, it does not separate individuals within any given species. Despite its name, the river is just a trickle in places and does not pose a physical obstacle to the migration of species (including *Homo sapiens*).

The noncoincidence of habitat and landownership means that the creation of property rights over genetic information will generate a number of owners for the same piece of genetic information. These owners will usually be on adjacent properties. Economists call such owners commoners—that is, people who hold something in common. In this case, the "commons"[3] is genetic information that is identical and stored in more than one place— that is, in more than one organism.

The delineation of a commons for genetic information is complex inasmuch as landowners may have valuable information in organisms on their parcel of land but not know who else has this same piece of information.[4] Most likely, they will also not have a clue as to its distribution over organisms. Recall that the storage of genetic information can be unique to any taxon from the family (the example given in Chapter 3 was Formicidae) to the genus (*Ursus*) to the species (*Zea diploperennis*) to the race (*Solanum melongena* 418 landraces) and even to the individual (John Moore). To delineate the commons for a piece of genetic information, one would have to first estimate its distribution across taxa and then determine the physical range of the organisms of the taxon in which the genetic information is distributed. Finally, one would have to identify all the landowners who have those organisms on their parcel of land.

To obtain income from the commons is even trickier. Landowners would have to convene and devise a mechanism to exclude anyone who has not paid some sort of user fee for the genetic information held in common. In sum, three types of transaction costs would be incurred: (1) the identification of the taxon at which the genetic information is distributed, (2) the identification of other landowners who have that same piece of genetic information on their land, and (3) the design and implementation of a scheme to exclude nonpaying users from access to that information. Clearly, these transaction costs are enormous. Because the costs of all three steps overwhelm individual landowners, each abandons the possible income that the delineation of genetic information would generate.[5]

How Much Money for a Landownership Database?

The answer for a small tropical country is hundreds of millions of dollars. Thailand is a case study.* Thailand has embarked on an exemplary 20-year project of issuing land titles to all freehold land parcels. The project is administered by the Department of Lands in Thailand and receives financial and technical support from the Royal Thai Government, the World Bank, and the Australian International Assistance Bureau.

The land-titling project will establish a national valuation system for land, produce a large-scale mapping, and improve the overall efficiency of land registration. The land-administration aspect alone entails 30 branch land offices, 10 new district land offices, 32 expansions and construction of 136 strong rooms in existing district land offices, and 102 houses for land office personnel. All this costs money and lots of it. The first five-year phase beginning in 1985 had a budget of $70 million in addition to the existent budget for the Department of Lands, which employs 10,000 individuals on a permanent basis. The second phase of the project costs an additional $100 million, with the largest emphasis on the development of a national land-information system.

*G. Feder, T. Onchan, Y. Chalamwong, and C. Hongladarom, *Land Policies and Farm Productivity in Thailand* (Washington, D.C.: Johns Hopkins University Press, A World Bank Research Publication, 1988); Ian P. Williamson, "Considerations in Assessing the Potential Success of a Cadastral Project in a Developing Country—A Case Study of the Thailand Land Titling Project" (Manuscript, Department of Surveying and Land Information, University of Melbourne, Parkville, Victoria, Australia, 1988).

Make no mistake. Just because the landowner abandons the potential income from a piece of genetic information does not mean that everyone abandons that potential income. Today the pharmaceutical, chemical, and agribusinesses of the North are obtaining income from the abandoned genetic information without paying anything to the commoners of the South. In the short run, this utilization of genetic information is highly profitable. Northern businessmen and consumers are enjoying the proverbial free lunch. However, the strategy is not tenable in the long run. In long run, "There is no free lunch," as Milton Friedman, the Nobel Memorial Laureate in economics loves to say. The long-run costs to the North will be the loss of GCFs as individual landowners in the South adopt alternative land uses that expunge genetic information.

For privatization to succeed as a conservation policy, the transaction costs of delineating genetic information and excluding nonpaying users from the commons must be drastically lowered. The only way to lower these costs is to capture economies of scale and reduce the average cost per landowner of establishing his share of the genetic-information commons. Theoretically, this can be done by following some simple steps to reduce the three types of transaction costs. Although the sequence of steps may be simple, each step is a complex task:

1. The identification of the taxon at which the genetic information is distributed will require molecular analysis of the organism for which the GCF has been commercialized and then molecular analysis of organisms from the same race, species, and genus to measure the distribution of that GCF across taxa.
2. The identification of other landowners who have that same piece of genetic information on their land will require not only a database of titles, but also a biological inventory with records for each land title.
3. The design and implementation of a scheme to exclude nonpaying users from that information will require that the industries that use natural genetic information in new products identify that usage, declare the dollar value of those products, and remit a royalty to the commoners.

Because genetic information is stored in organisms and organisms migrate, the biological inventory will also require continual data input. Likewise, titles change hands, and the database of titles will require continual revision. And finally, the commercialization of natural genetic information is an ongoing process. Consequently, the match of genetic information, landowners, and industrial usage is very dynamic. The merging of these databases will be a gargantuan database, which will also be very dynamic.

The gargantuan database delineates the commons. Making the match will be costly, inasmuch as each database may cost billions of dollars. Just the first two steps will require government intervention on an immense scale. In other words, a bureau and bureaucrats.

The suggestion of a new bureau may raise the hackles of free-market purists. After all, in these times of government retrenchment, what is the rationale for a new multibillion-dollar bureaucracy? Cannot entrepreneurs perform these tasks? Doesn't more government contradict the very spirit of privatization? To answer the free-market purists, one must consider again the peculiar economics of information, already mentioned briefly in the analogy with software in Chapter 4.

Information has a property unlike any other commodity: it is costly to create but almost costless to reproduce. Or, in the rhetoric of mainstream economics, the fixed costs are very high and the marginal costs are almost zero for the first producer as well as all other producers. This means that if government lets the market create a gargantuan database, then later entrants to the market will either incur the huge fixed costs of reinventing the wheel—steps (1) and (2)—or they will free ride on the information gathered by the pioneering firm. Government intervention eliminates the duplication of fixed costs and obviates the free-rider problem.

Another compelling reason for government intervention is the transaction costs of communication.[6] If there are several smaller, private databases, all using their own protocols, commoners will have difficulty communicating among one another. To appreciate the costs of communicating without protocols, consider the following analogy: *one gargantuan database of genetic information and landownership* is to *several mini–gargantuan databases* what *one currency* is to *many currencies*. Just as there can be several currencies used in international trade, there can be several bureaus of database management to track genetic information and landownership. And just as the transaction costs of international trade are greatly reduced if we adopt one currency and avoid the hassles of foreign exchange, the merging of huge databases on genetic information and landownership will also be facilitated if each commoner adopts the same protocols in communication networks. Such protocols enable efficient electronic data interchange (EDI) and thereby reduce the costs of delineation.

Although privatization as a conservation policy cannot proceed efficiently without some degree of centralization, centralization does not have to mean that the bureau will engage in every aspect of data collection and analysis. I have chosen the analogy with foreign currencies carefully. Although a central bank is commissioned to govern the money supply, its commission does not include setting prices on goods and services. And just as the public tolerates the inefficiencies of a central bank, the public should be willing to tolerate some inefficiencies with the bureau that manages the gargantuan database.[7]

Because the gargantuan database is crucial in delineating the commons, its institutional design and implementation are of paramount importance. The complexities of the database are so overwhelming that, to use an apt metaphor, I will present my vision of the forest through the trees. Where it is basically correct, experts in a variety of fields will have to bring the trees into focus. Where it is basically wrong, experts will have to suggest alternative vantage points, new forests, different trees.

To understand how the gargantuan database will handle the biological inventory, one must first understand how biologists presently handle similar biological inventories. The starting point is taxonomy, a science and art essentially unchanged since the eighteenth century, when Swiss naturalist Carl von Linné standardized it. A good example of twentieth-century Linnaean taxonomy can be gleaned from an article, "Could There Be a New Species of Animal in Your Backyard? Putting the Principles of Classification in Context." The author, S. A. Marshall of the University of Guelph, begins by telling us that it is quite possible that there is a new species of insect in the North American backyard. Then he goes through the tortuous path of establishing whether or not an insect in your backyard is indeed a new species. He concludes:

> At a minimum, to describe a new species you must give it a Latinized species name which is different from other names in the genus, you must write a description, you must designate a type specimen and you must get your paper including the above information published. Once it is published, your species isn't new any more. It is a species that you described.[8]

One is not quite sure whether Marshall's intent is to encourage or to discourage amateur taxonomists. I know I was discouraged! Listing, cataloguing, and describing a species and then making a definitive judgment on its place in the tree of life cannot be done in an afternoon. Sometimes it takes years, and even then taxonomists will disagree over whether or not an organism belongs to a new species or is simply a variety of the same species. Given that less than 10 percent of the biota on the planet has been classified in the manner Marshall describes, what would it take to classify the remaining 90 percent? A simple calculation can shed some light on the answer. Assume that there are 25 million unclassified species and that it takes a taxonomist one year to classify 25 species. That computes to 1 million taxonomist years. By many estimates, there are only 1,000 taxonomists actively engaged in classification. So it would take 1,000 years to classify the planet. Say the average taxonomist requires $30,000 per year. The cost of classifying the planet at the species level is then $30 billion. Even if taxonomists were to find the time and even if governments were to release the money, the funding would in no way guarantee the survival of the species once classified. Taxonomists may just end up classifying newly extinct species.

Because the delineation of the genetic commons requires taxonomy and exhaustive inventories, the traditional taxonomy that Marshall describes is inadequate for privatization as a conservation policy. Ways must be found to speed up the classification and count of specimens.[9] This does not have to mean a cursory biological survey in lieu of exhaustive studies and elaborate phylogenies.[10] Rather, it can mean routinization of tasks that complement rather than substitute for the scientific rigor of traditional taxonomy. This is a daunting task, inasmuch as the routines must also complement routines involved in the search for land titles of habitats.

How to begin routinization? One need not be an expert to surmise the answer: computer management of information. However, one will need a great deal of expertise to design the system of routines. The system that occurs to me is pattern analysis of photographed specimens. In the pages to follow, I will suggest some of the routines; admittedly this is an amateur's attempt to see the forest and some of the tallest trees. Here goes.

Imagine that genesteaders take photographs of representatives of an unusual organism on 35mm film according to protocols established at the bureau of the gargantuan database. The protocols for plants may include, say, the buds, flowers, and seeds; for mammals, the dentition, ears, and genitalia. The photograph will then be "digitized"[11] and recorded on an electronic map—that is, a geographic information system (GIS)—under the record of the owner of the land where the organism was found. Each photograph then receives an alphanumeric binomial voucher. For example, the genestead owned by Juan may be the code A5ZN2BI5. The code becomes a directory for digitized photographs on Juan's genestead. This means that bureaucrats at the gargantuan database can pull up on screen A5ZN2BI5 and find information on the owner (e.g., national identity card) as well as all the photographs Juan has submitted, now digitized images. Each of these images will then be given a sequential number. The first organism photographed from genestead A5ZN2BI5 might be A5ZN2BI502-1f, where "f" denotes an image of the flower for organism "1"; the hundredth organism photographed might be A5ZN2BI502-100g, where "g" denotes an image of the genitalia of that organism.

Inasmuch as many organisms are morphologically similar, Juan will photograph only a representative of the group. Juan is no biologist. Undoubtedly, he will make mistakes. For example, he may think the drab female of a species is an organism of a different group than the ornate male of the same species. He may submit many redundant photographs. So what? No system is foolproof. As Juan learns more about the ecology of his genestead, he will discover his mistakes and correct them through communication to the bureau of the gargantuan database.

At the bureau, the digitized images can then be recorded on a write-once-read-many (WORM) optical disk system. WORM disks can store huge volumes of multimedia information and achieve the resolution of 35mm film.

From Here to There

The delineation of the genetic commons relies heavily on photography and artificial intelligence. I have presented the proposal to at least a dozen taxonomists and all have objected, some quite vehemently. They object on two counts:

1. Photography is no substitute for specimens.
2. Artificial intelligence is no substitute for taxonomists.

Apparently, state-of-the-art photography is not sufficiently developed to pick up the nuances of specimens from different families, much less from different races. Moreover, the application of artificial intelligence to photography is still in its infancy. All the taxonomists seem to agree that the system I propose will not work today. I have no choice but to accept their expert opinion; nevertheless, I believe that they have no choice but to accept that the wave of the future is artificial intelligence. Across every sector of the information economy, computers are becoming substitutes for labor.

Although most taxonomists disagree with photography as a definitive classification tool, most would agree with the idea of royalties. Many suggest that the royalties go to a general environmental fund for the country for which the plant is endemic. Such a suggestion goes against the efficiency criterion of property-rights analysis: *those who control an asset should be the ones who derive the benefits from that asset.* If royalties go to a general fund, then the people who are generating the benefit (the landowners who do not deforest) will not be receiving the full benefits they generate. It is even conceivable that they will not receive anything, despite the collection of royalties for the genetic information they control. The money could easily be diverted to other government programs.

The scheme presented in this chapter is a way to allow individual landowners to claim the royalties. But because the appropriate technologies in photography and artificial intelligence are not yet available, the taxonomists are right: there is no practical alternative but to place the royalties in some type of general environmental fund. The fund should, as best it can, remit part of the royalties to the landowners of habitat typical for the species commercialized, and the other part should be set aside for research on the application of artificial intelligence to taxonomy. As photography and artificial intelligence reach a level where each landowner can classify the organisms on his land accurately and cheaply, then the general environmental fund should give way to the gargantuan database.

Each WORM disk can hold roughly 36,000 images. Because records on WORM cannot be purposely or accidentally altered, WORM technology can reduce the negotiation costs of disputes. Say, for example, a plant with a useful alkaloid is discovered on the genestead of Dolores, Juan's neighbor; the royalties may be worth hundreds of thousands of dollars. Juan gets wind of Dolores's good fortune. He searches his genestead for the same organism, but to no avail. Under cover of darkness, Juan sneaks over to Dolores's genestead and steals a couple of plants. He starts cultivating them, takes a few photographs, and then submits the photographs to the bureau.

A Crafoord Is Not a Nobel: Recognizing Biologists

Privatization will give new respectability to one of the oldest of professions: taxonomy. There is a malaise among young and old taxonomists for a very simple reason: no money and little recognition. In the United States, young biologists cannot even get poorly paid assistant professorships. Norman Carlin, a Harvard Ph.D. in evolutionary biology (1986), reported his inability to find an academic post after five years of trying; he has given up and is retraining in the law.* Tenured biologists in academe are also discouraged; the highest levels of achievement go unrecognized. Indeed, there is no Nobel Prize in organismic biology.

Perceiving this situation, the Royal Swedish Academy of Sciences created the Crafoord Prize. Biodiversity advocates like E. O. Wilson and Paul Ehrlich have received the award. However, Crafoord is not a household word like Nobel. As such it does not carry the prestige among the powers that be. For example, heads of state will often cite the advice of the laureates in economics to legitimize ideologically driven policies. Yet there is no Nobel Prize in economics: But what about Samuelson, Friedman, and the 1991 laureate, property-rights analyst Ronald Coase? Aren't these Nobel laureates in economics? The answer is no. They received a prize that is *in memoriam* to Nobel, but the prize is not a Nobel Prize. The Royal Bank of Sweden created it in the late 1960s. So the question arises: Why doesn't the Royal Swedish Academy abandon the Crafoord and use the money to create another Nobel? In the politics of persuasion, there is nothing like a Nobel, even if a Nobel is not a Nobel. If economists can appropriate the title "Nobel," why can't the biologists?

*Leon Jaroff, "Crisis in the Labs," *Time*, 26 August 1991, 45–51; Jon R. Luoma, "Taxonomy, Lacking in Prestige, May Be Nearing a Renaissance," *New York Times*, 10 December 1991, C4; "Britain Funds Taxonomy Studies," *Nature*, 1 April 1993, 383; Peter T. Kilborn, "The Ph.D.'s Are Here, But the Lab Isn't Hiring," *New York Times*, 18 July 1993, E3.

Juan's theft will not work. The trouble is that WORM technology does not allow Juan to pre-date his records for A5ZN2BI502. The newly submitted photograph will be digitized and entered as A5ZN2BI502-101f. In the record for the organism 101 is the date of submission for the photograph, which is after the filing date of the patent and notification of commoners like Dolores. Juan cannot rig his claim as a commoner because WORM enables an audit trail. If Juan's inventory is incomplete, then he is out of luck. Like many patent systems based on the first-to-file principle, this policy is based on the inventory at the time of patent. This may seem unfair; however, it is less unfair than the alternative, which is rampant piracy.

How to begin classification of the digitized photographs? Although computers are a substitute of capital (the computer) for labor (the traditional taxonomist), the amount of labor (the supply of traditional taxonomists) required would still be phenomenal. The job is immense and extends beyond the 25 million species unclassified to include subspecies. In other words, despite labor-saving computerization, there would still not be enough professional taxonomists to meet the demand. A labor supply must be quickly found.

Fortunately, there is a reserve army of underpaid and understimulated individuals who could not only do the job well but probably also enjoy doing it. They are the secondary-school teachers of biology. The classification of species can take the following form. Each school in the North would enter into partnership with a genestead in the South (the financial implications to be discussed in Chapter 9). The biology teacher assists in the classification of the species by examining the digitized photographs submitted by the genesteaders. The examination is assisted by a computer program designed by experts in artificial intelligence and taxonomy. One such system that comes to mind is *Linnaeus*, an interactive taxonomy using the Macintosh computer and HyperCard database, also known as the HyperCard stack.[12] The HyperCard stack contains the taxonomic keys that allow one to ferret out the relationship of a particular organism among similar-looking organisms.

As knowledge of a tropical ecosystem accumulates for a particular region, the HyperCard stack is upgraded by the gargantuan database. The biology teacher who assists in accurately classifying a species would then receive some percentage of royalties should something commercial result from the species classified. Inasmuch as secondary-school biology teachers are notoriously underpaid, their response to this incentive should be strong.[13]

The value for education is incalculable. Recall from the discussion in Chapter 2 that all the great conservationists see education as the long-term solution. A spin-off of the policy is that it affords educational opportunities for the young, the very ones who have the most at stake in the mass-extinction crisis. Moreover these opportunities are not the deadpan pedagogy so typical of state-run education. Students will be actively engaged in scientific discovery as they become sensitized to an environmental ethic. Ideally, an exchange of expertise would take place with biology teachers from the North spending a summer in the South. During their visit, the teachers could transfer the latest knowledge about habitat management. The transfer of knowledge could even extend beyond habitat management to personal health, sound business practices, and human rights.

To make the final classification, the biology teacher downloads the information on the species and communicates it through one of the electronic bulletin boards in taxonomy. One of the first such services is TAXACOM, of the Clinton Herbarium of the Buffalo Museum of Science in Buffalo, New York.[14] Although the service is free, the school must pay for the long-distance phone charges. Phone charges can quickly accumulate if the work on the expert system is done on line. This is why the system should be decentralized on micro computers. Students learn on the micro computers, and then their teachers refine the classification and submit it to the experts via TAXACOM, who assist them in composing Latin descriptions of the new taxa.

This is not a plug for either TAXACOM or *Linnaeus* (although they may indeed be excellent programs and very deserving). What I am plugging is that we digitize the photographs through a WORM technology, that the digitized photographs be recorded with landownership on a geographic information system (GIS), that the preliminary classification of the digitized photographs be done on micro computers by secondary-school biology teachers, that the classification then be communicated online for critical comment by one of the precious few experts, and that the gargantuan database upgrade the classification programs on the micro computers as species are definitively classified.

Now let us say that while all this is going on, Merck & Co. receives approval from the U.S. Food and Drug Administration for a new drug whose active ingredient comes from a tropical plant in one of its hothouses. Who will get the royalties? The logical starting point would be to ask Merck & Co. where it obtained the plant, go to that source, and survey the habitat of the species. Suppose it is not as simple as that. In the Third World, nothing is ever simple. Merck may have paid a plant collector to bring back specimens, and the plant collector may be wary of divulging the habitat for fear of being cut out of his collection fees. All Merck may have are some hothouse plants and a closed-mouth collector. If that is all there is, well then that is all there is. One makes do.

Merck will photograph the specimens according to the same protocols used by the genesteaders. Once Merck submits the photographs, the match can begin. The gargantuan database digitizes the photographs and, through pattern analysis of the buds, flowers, and seeds, locates all the genesteads that have submitted photographs with similar-looking buds, flowers, and seeds. Imagine that after whirring for some minutes, the computer at the gargantuan database displays the voucher numbers A5ZN2BI50-86b, A5ZN2BI50-86f, A5ZN2BI50-86s, X2N9C43-1356b, X2N9C43-1356f, X2N9C43-1356s, C8P2RY21-634212b, C8P2RY21-634212f, and C8P2RY21-634212s. These numbers mean that the buds, flowers, and seeds of the 86th organism photographed on genestead A5ZN2BI50, the 1356th organism on genestead X2N9C43, and the 6342nd organism on genestead C8P2RY21 all match the images of the Merck hothouse specimen. The species seems to be endemic to a small region of the tropics that make up just a few genesteads.

Unfortunately, the match described is not definitive. For example, the marsupial mouse looks just like a placental mouse, even though it is more closely related to a kangaroo! This is also true with plants. Many trees in the Amazon are from totally different taxonomic families yet have almost identical leaves. Digitized images can be deceiving. Therefore, before Merck disburses any royalties, some sort of rigorous taxonomy must be deployed.

To ascertain the true distribution of the same piece of genetic information, each of the genesteaders identified through the match is contacted. The genesteader is asked to submit a sample of the organism photographed. Tissue from the sample undergoes molecular analysis.[15] Through DNA sequencing, molecular biologists can determine whether the sample contains the same genetic information that coded for the active ingredient in the drug. Tissues are then analyzed from different races within the same species, and then from different species within the same genus, different genera within the same family, and so on, until the distribution of the genetic information over taxa is ascertained. With the identification of both the taxon and the genesteaders who have organisms from that taxon on their land, the genetic commons is delineated. The success of the policy now turns on the design and implementation of a scheme to exclude nonpaying users from access to the genetic-information commons. Such a scheme takes us into the complex world of politics and finance.

8

Politics

Exclusion of nonpaying users cannot be divorced from finance, and finance cannot be divorced from contracts, and contracts cannot be divorced from the law. Therefore, to talk about the exclusion of nonpaying users, one must also talk about the law. The laws of interest are those governing patents, copyrights, semiconductors, trademarks, and trade secrets. These are the laws that define intellectual-property rights. These rights, in turn, set the parameters for contracts between creators and users of intellectual property. Because the laws protecting intellectual property are drafted by politicians, exclusion ultimately cannot be divorced from politics.

A word of caution: the legal meaning of "property right" in the term "intellectual property right" is not the same as the economic meaning subscribed to in previous chapters. Recall that in economics, "Property rights of individuals over assets consist of the rights, or the powers, to consume, obtain income from, and alienate these assets." For example, you could photocopy this book and hawk it on the streets of Jakarta. In the economic sense, you would have a property right, inasmuch as you are physically able to make a photocopy—you have "the powers to consume, obtain income from" the photocopy. However, this does not mean that you have a property right in the legal sense, in which a right is defined as the "capacity residing in one man of controlling, with the assent and assistance of the state, the actions of others." You do not have the assent and assistance of the state. Indonesia is a signatory to international agreements on copyrights. These agreements deny you the intellectual-property right. But to the extent that Indonesia does not enforce its copyright law, you can nevertheless enjoy an economic property right of consuming and obtaining income from my intellectual property—that is to say, this book.

The connection between the two meanings of property rights lies in the transaction costs of policing. As countries like Indonesia begin to respect intellectual-property rights and enforce the law, the transaction costs in policing decrease and, hence, my economic property right "to consume, obtain income from" the book increases. But the property right is never fully delineated. Even with the assistance of the state, my policing costs will, at some point, be greater than the value usurped by the hawkers. At that point, I abandon my economic property right, even though I have never surrendered my legal property right.

In earlier chapters, it was argued that landowners have abandoned their economic property right over genetic information due to the high policing and negotiation costs of excluding collectors and delineating the commons. A gargantuan database was suggested to reduce the negotiation costs. To reduce the policing costs of excluding nonpaying users requires legislation of equal protection of artificial and natural information (first introduced in Chapter 5) and new institutions to enforce the new legislation.

A law of equal protection would reduce the policing costs at the point of patent. The pharmaceutical companies, chemical industry, and agribusinesses that now "consume, obtain income from" genetic information would have to reveal to the gargantuan database (1) the natural source of the information or (2) the logical steps that led to the synthetic creation of complex biomolecules. If the latter is a deception, then the employee will have committed fraud, a criminal offense. Because the risk of being caught would be born chiefly by an employee but the benefits would be distributed over the whole company, individual employees would have little incentive to deceive.

Legislation would also reduce the negotiation costs between the commoners of genetic information and the industrial users by setting the parameters of the contract. In other words, there would be no haggling over the duration of the right or the percentage of the royalty. These and other parameters would be embedded in the legislation. The law would become the standard for the contractual relationship between the genesteader and the commercial user.

In effect, equal protection of artificial and natural information is nothing more than an extension of existent intellectual-property law. Countries often extend the domain of intellectual property when it is clearly in their interest. For example, the United States drafted the Chip Protection Act for semiconductors in 1984 to protect the burgeoning computer industry, and Australia drafted the Plant Varieties Act in 1987 to protect genetically altered plants.[1] Both made sense, and both were in the best interests of the respective country. New legislation of equal protection would be the natural progression of intellectual-property rights for countries that, on a per capita basis, are genetically rich. Except for Australia and New Zealand, most of these countries are in the economic South.

It is now in both the short- and long-term interests of the South to privatize genetic information and pass such legislation. There is no economic reason why any Southerner should object. However, the same cannot be said of the North. The North may resist because it will have to pay for something it is used to getting for free. In the short run, the pharmaceutical, chemical, and agricultural industries will be adversely affected, inasmuch as the costs of royalties cannot be completely passed on to the consumer. Management understands this basic economic tenet. However, if management really learned their economics well, they will recall that the extent to which they can pass on the costs is a function of the sensitivity of demand for the good, what economists call price elasticity.[2]

Two examples may show how price elasticity will determine which industries will bear the heaviest and lightest burdens of royalties, should genetic information be privatized and, by the assumption of selfish interests, who will also be the most vociferous in their opposition to equal protection. The first example is *Diploglottis campbellii*, an ornamental tree with large edible fruits. The second is that old favorite of conservationists, *Catharanthus roseus*, the rosy periwinkle.

Let us consider the fruit tree first. The species *Diploglottis campbellii*[3] is endemic to Queensland, Australia. There are only seven known trees. All of them are on private property. If agribusiness starts cultivating orchards of *Diploglottis campbellii* for commercialization of the large edible fruits, then agribusiness, under the proposed legislation, will have to remit 15 percent of gross receipts to the three proprietors in proportion to how many trees are on their property (for example 1/7, 2/7, and 4/7) for 17 years. Inasmuch as demand for fruit is highly elastic (if pears become expensive, then people eat more apples and fewer pears), agribusiness will not be able to pass on the royalty costs to the consumer. They will have to absorb that cost.

Any policy that raises the costs of doing business will quickly be felt in the pocketbooks of management, whose remuneration is ultimately tied to profits. It therefore becomes in the selfish interests of managers in agribusiness to oppose privatization. Although the policy may sound like the conservative rhetoric they fancy, they may still shake their heads and argue that this is not free-market economics but a socialism cloaked in the rhetoric of free markets, a ruse to redistribute wealth, and a shakedown of the North by the South.

There should be less opposition and eventual support of privatization by the pharmaceutical industry. This is not because the pharmaceutical industry is enlightened.[4] Hardly. The reason is selfish interests. To perceive these interests, consider again the two chemotherapeutic drugs derived from the rosy periwinkle of Madagascar, vincristine and vinblastine. Unlike the large edible fruits of the tree *Diploglottis campbellii*, there is no substitute chemotherapy for Hodgkin's disease and juvenile leukemia. In the rhetoric of economics, the demand for the fruit is elastic, while the demand for vincristine and vinblastine is inelastic. It is this elasticity that determines what share of the 15 percent royalty cost gets passed on to the consumer. For the fruit of *Diploglottis campbellii*, the market may clear with the consumer picking up another 1 or 2 percent of the royalty costs and agribusiness picking up 13 or 14 percent. For vincristine and vinblastine, it will probably be just the opposite, with the consumer picking up 13 or 14 percent and the pharmaceutical industry just 1 or 2 percent.

A decline of 1 or 2 percent is still a significant loss; why would the pharmaceutical industry support privatization? There are two answers to this question: selfish interests in the long run and selfish interests in the short run. As elaborated in Chapter 4, privatization will enhance efficiency in the long run as genetic information is preserved and traded. However, it is doubtful that the CEOs of any industry think in the long run. Unfortunately, in a competitive market for capital, CEOs are hostage to stockholders worried about the bottom line in the shortest of short runs—the quarterly term of financial reports. These CEOs know too well that in the long run there will be new CEOs who not only would reap the benefits of their sacrifices, but, to add insult to injury, would also take credit for creating those benefits!

The only short-run return for CEOs is rhetoric. The Madison Avenue potential of an industry imposing a cost on itself in order to save biodiversity is a rhetorical bonanza, one the pharmaceutical industry desperately needs. Ironically, the industry needs good rhetoric to offset the bad rhetoric of its own capitalistic success! Over the past 10 years, pharmaceutical houses have been exacting the monopoly price whenever they can get away with it. In an otherwise flat world economy, the industry thrives and its stock prices have soared. The story of Burroughs Wellcome and the drug AZT is one example, albeit an extreme one, that illustrates how capitalistic success engenders a bad rhetoric and ultimately threatens profitability.[5]

AZT is effective in repressing the replication of HIV, the AIDS virus. AZT is an old chemotherapeutic drug that found a new indication with the advent of HIV. Obviously, HIV+ individuals are willing to pay dearly for an extension of life; the demand for the drug is inelastic at prices comparable to drug therapies for other fatal diseases. Burroughs Wellcome holds the patent on AZT and understands, all too well, the microeconomic concept of inelasticity. It has raised the price of the pills to $2000 to $3000 per year and now reaps windfall profits. However, AIDS activists keep the story of price gouging alive—bad rhetoric not only for Burroughs Wellcome, but also for the whole industry. Indeed, activists are calling for regulation of pharmaceutical pricing.[6] These calls are more likely to fall on deaf ears if the industry can demonstrate acts of public citizenship like saving the rain forests. "Royalties to save rain forests" may serve these rhetorical needs.

There is also another reason why the pharmaceutical industry may be receptive and agribusiness resistant: the relative dependence on biodiversity. Despite the environmentalist's rhetoric about 25 percent of all drugs originating in the tropics, research laboratories are in fact very independent of biodiversity in their research. Scientists have largely ignored the potential of plant sources and prefer to computer manipulate existing molecules for marginal patent improvements.[7] This practice is only beginning to change.[8] So, to the pharmaceutical industry, the royalty costs of privatization as a conservation policy would be a hypothetical cost, something out there, while the rhetorical bonanza would be here and now. This is not the case with agribusiness. All the common foodstuffs are heavily dependent on infusion of virus-resistant genes from land races of domesticated crops. From these land races, new hybrids are commercialized. So it is in the selfish interests of these two industries to see privatization quite differently.

By considering the pivotal role of rhetoric in decision making, the preceding analysis departs radically from standard textbook economics. Probably better than any other sector of the world economy, agribusiness knows how legitimate such a departure is. Public policy on agriculture has almost nothing to do with textbook economics. The policies are usually driven by a rhetoric that, upon scrutiny, reveals vested interests. Therefore, one need not be a cynic to suspect that agricultural interests will pay their lobbies handsomely to manipulate the rhetoric and politick against privatization as a conservation policy.

This may sound too cynical and pessimistic. Let us be optimistic and suppose that legislation giving equal protection to artificial and natural information passes in the United States. Despite the enactment, the new law will still be in jeopardy. Agribusiness will quickly seek legal recourse. Lobbies like the American Seed Trade Association will hire top-flight attorneys to oppose such legislation as unconstitutional. It takes only one well-funded

Rhetoric and Vested Interests

One cannot assume that the North will be ideologically consistent if such consistency goes against vested interests. For example, the United States prides itself as the bastion of freedom, and yet in the 1970s the United States allied itself with the murderous regimes of the South by lending billions of dollars to them. What the dictators did not squander on foolish projects they stole outright. The debt is now unpayable. A well-known consequence of the Third World debt is deforestation. Had the United States been consistent in its ideology of freedom, it would never have recognized the legitimacy of dictators to contract public debts. Such consistency would have averted the debt crisis, fomented democratization, and at least postponed deforestation and the ensuing mass extinction.

That was then. This is now. Yet nothing has changed! Sound economics and ideological consistency are still very much subordinate to vested interests disguised in noble rhetoric. In 1990 the governments of the North cooperated hand-in-glove with Saudi Arabia to "free" Kuwait when Kuwait had never been free. To free Kuwait would be noble if those who were participating were also free. However, it was never suggested that Saudi Arabia honor human rights (e.g., the penalty for conversion to Christianity in that country is death). So the mere fact that privatization as a conservation policy matches the rhetoric of the North is not enough. Just as vested interests wanted to "free" Kuwait, vested interests will want to keep genetic information "the common heritage of mankind" even if "the common heritage of mankind" means a glaring contradiction in the ideology they profess.

lobby to challenge the legislation, and most patent attorneys would probably doubt its constitutionality.

If equal protection is ruled unconstitutional, we are back at square one with the prohibitive transaction costs in policing and excluding nonpaying users from the genetic-information commons. Although there is always the possibility of a constitutional amendment, constitutional amendments move at glacial speed. Something as simple and just as the Equal Rights Amendment (ERA)[9] has languished for decades in the United States. If the same were to eventuate for legislation of equal protection, preservation will become a moot issue—the useful genetic information will have been expunged. Therefore, advocates of privatization must, at the outset, anticipate the legal challenge and be prepared with a legal argument for equal protection as well as an alternative tactic should that legal argument fail.

A legal defense of equal protection càn be developed by marrying original intent to information theory. Original intent refers to the motivations of the drafters of the United States Constitution: How would they interpret novel events with the reasoning embodied in the constitution? This is a conservative ideology eloquently expressed in the writings of the failed Supreme Court nominee Robert Bork. It can easily be argued that the original intent of the drafters of the Constitution corresponds to the equal protection of artificial and natural information. To do so requires techniques from modern literary analysis.

Literary scholars may laugh at such an initiative. To them, the doctrine of original intent is anachronistic in this age of deconstruction and semiotics.[10] The New Age theorists argue that it is impossible to divine original intent: the sender and the receiver simultaneously define the message by their actions and reactions. Once the sender and receiver are dead, the message may linger but it will take on new meanings for new purposes and new occasions—it is not the same message. To some extent, this is of course true. Religion is the prime example. What Jesus said in Aramaic to his disciples has been evolving over the centuries. Today in Latin America, the Gospel has become a call to arms and Jesus an Uzi-swinging Sandinista! Through the lens of semiotics, Jesus' original intent is unattainable and irrelevant. But a partial truth is not the whole truth. Although liberation theologians can reconstruct the message Jesus sent for new purposes and new occasions, the emotions that yearn for leadership and justice are probably the same across the centuries and millennia. In other words, to the extent that a message is in tune with basic human behavior, the message will still trigger similar actions and reactions by senders and receivers long after both the original senders and receivers are dead.

The same should hold true for what the drafters of the Constitution wrote. To discover their original intent, one must consider both the language that deals with intellectual property and the context in which that language was written. The language can be found in Article I, Section 8, of the Constitution: "To promote the progress of science and useful arts, by securing for limited times to authors and inventors the exclusive right to their respective writing and discoveries." To appreciate the context of that language, one must consider the environment in which the founding fathers lived: a frontier, hardly explored, much less exploited. To discover their original intent, one must reduce the language into a more neutral terminology that can be placed into today's context—that is, an environment of degraded ecosystems and mass extinction.

To reduce Article I, Section 8, one must examine each phrase. "To promote the progress of science and useful arts." This language is clear; few would disagree that this means the provision of incentives. "By securing for limited times" implies that the incentive—a monopoly right in the new art—will be surrendered after a fixed period of time. The intent here is quite noble: eventually to distribute the gain of the invention to the public. "To authors and inventors the exclusive right to their respective writing and discoveries." It is in this phrase that legal scholars flounder. They equivocate on the words "inventor," "writing," and "discoveries" by slavishly adhering to a literal interpretation of the words when such adherence may not have been the founding fathers' intent had they not lived in a frontier. From there the Court infers a work-based property right. For lack of a better term, I will call this ethic *compensation for work expended.*

There are many examples where the Court has inferred compensation for work expended. One that has special relevance to genetic information is the landmark 5 to 4 decision granting a patent to a live organism.[11] In 1980, the Supreme Court upheld the patentability of a DNA-engineered organism: "Here by contrast, the patentee has produced a new bacterium with markedly different characteristics from any found in nature and one having the potential for significant utility. His discovery is not nature's handiwork, but his own; accordingly it is patentable subject matter under [the basic patent law.]" Or conversely, if it is nature's handiwork, it is up for grabs. Compensation is only for work expended.

This was not the original intent of "to authors and inventors the exclusive right to their respective writing and discoveries." Had it been, the founding fathers would have certainly disallowed inheritance. They did not, even as it pertained to slaves. The original intent of Article I, Section 8, is merely to provide incentives, nothing more. This becomes obvious when one reduces the wording all the way down to the immutable level—the physical essence of "inventor," "writing," and "discoveries." The physical essence of any process, manufacture, or invention consists of signals that convey to the receiver the mechanism of the process, manufacture, or invention. By the Shannon–Weaver equation of information theory, these signals can be reduced to a sequence, and the probability of that sequence can be quantified as bits of information. Inasmuch as a process, manufacture, or invention is not the product of natural forces but of human will, the information is artificial. Hence, the "writing" or "discovery" is artificial information, and *the inventor controls its existence.*

One can reduce natural phenomenon similarly. In the case of the rosy periwinkle, the physical essence is a sequence of base pairs that likewise can be reduced to probabilities and bits of information. Hence, the rosy periwinkle is also information—natural information. Although the creator of the rosy periwinkle is not an individual but a process (evolution), landowners forgo opportunities to preserve that rosy periwinkle. So, *the landowner controls its existence.*

Whereas the inventor of a machine can enjoy a property right over the artificial information embodied in his machine, by "the common heritage of mankind" the owner of the habitat for the rosy periwinkle cannot enjoy any property right over the natural information embodied in the periwinkle. Why should preservers of natural and creators of artificial information enjoy equal legal protection? The answer is the original intent embodied in Article I, Section 8: "to promote the progress of science and useful arts." Here the progress are the two chemotherapeutics, vincristine and vinblastine, whose sales have passed $1 billion. Without ownership over such natural information, there is no incentive for the one who controls the existence of the periwinkle "to promote the progress of science and useful arts." Inasmuch as it takes incentives to preserve natural information just as it takes incentives to create artificial information, the twentieth-century word "preserver" and the eighteenth-century word "inventor" can be equated as "controllers over existence." Such an interpretation would render the new legislation constitutional.

If the Supreme Court is not persuaded by this line of reasoning, all is not lost. A concerted effort should then be made by a coalition of groups to pressure the government for a constitutional amendment. The coalition should extend beyond the usual environmental activists and include articulate politicians of the South. In other words, Southerners should take an active role in northern politics. This contingent plan is essential, inasmuch as *legislation is crucial to reduce the transaction costs of policing and excluding nonpaying users from the genetic information commons.*

The attack should take two forms: (1) rhetorical threats to northern politicians who refuse to support equal protection, and (2) economic threats to northern industries that continue to free ride on those who preserve genetic information. In the attack, Southerners should dispense with meekness and humility. Polite diplomacy is inappropriate when nearly 1 billion Southerners are in clinical starvation. Privatization could save many of them by providing a livelihood. Now is the time for the South to assert itself. The following two vignettes will illustrate how this can be done.

The first vignette is set in Brazil in 1991. The Bush administration sends Vice President Dan Quayle with an entourage of American industrialists to Brazil. The visit has one purpose and one purpose only: to pressure the Brazilian government to reform intellectual-property law. Clearly, much U.S.

intellectual property is being pirated by Brazil, and any enforcement of international law would mean a transference of money from Brazil to the United States. The American industrialists, however, are viewed by the Brazilian public as a mafia and Quayle as their lackey. With his usual adroitness, Quayle spells out what will be the economic penalty if Brazil does not submit: no agreement on transfer of technology. The president of Brazil, Fernando Collor, recognizes that "no transfer of technology" is an American euphemism for "no new IMF/World Bank credits." In desperate need of new credit, Collor endorses intellectual-property reform in Brazil. Less than a month later, a full-page article appears in the *Jornal do Brasil* (*JB*) entitled "Bush Promises to Help Brasil." Two days after that, a front-page headline announces, "Collor will Ask Bush for Foreign Investment," and the figure of $1.5 billion is cited. The money will come from the IMF. Finally on 24 September 1991, we see on the front page of the *JB*, and above the fold, Bush seated with Collor in the Waldorf Astoria Hotel in New York City. The caption reads "Bush Promises to Do 'Everything Possible' to Help Brazil with the Renegotiation of the Foreign Debt."

Although the Bush–Quayle diplomacy was incredibly crude, America does have every right to flex its muscle, and Brazilians should respect American intellectual property. Now it is Brazil's turn to flex its muscle and be just as crude, just as blunt, and hopefully just as effective. The vice presidents of Brazil and other countries of the South should go North and say simply, reform *your* constitution as we are reforming ours *or else* we are left with no other choice but to burn the forests. Rather than being ashamed of the burnings, they should televise the burnings with the faces and names of American politicians who are against the constitutional amendment. Below the name of each politician should be the dollar sum received in campaign contributions or speaking fees from the vested interests opposed to the legislation. No politician in the North wants to be perceived as mercenary, anti-environmental, and pro-starvation. They will cave in.

The irascible governor of the Amazons, Gilberto Mestrinho, has suggested in jest something very similar:

> Every year you have American and foreign politicians telling us what we can and can't do with the Amazon. Can you imagine a group of Brazilian Senators telling President Bush what to do with the forests in California or Alaska? The Amazon is not a world monument. It belongs to Brazil, and it is up to us to decide its future.[12]

So just as Americans traveled south to tell Brazilians how to reform the Brazilian constitution, now is the time for Brazilian leaders to travel north and tell the Americans how to reform the U.S. Constitution. What is good for the goose is good for the gringo.

The second vignette will illustrate how to hit hard the industries that bankroll the lobbies that co-opt the members of Congress who might vote against a constitutional amendment to mandate equal protection of natural information. The key is competition. The South must foment a "gene rush" for the newly created property rights over genetic information. Here is how the scenario can unfold. In the biodiversity game, Brazil is the major player. In finance, it is Japan.[13] The two countries have strong cultural links through waves of migration that span generations. Because of these links, Brazilians can play off the Japanese against the Americans and Europeans to respect equal protection of artificial and natural information.

This vignette is set in the state of Acre, on the Brazil–Peru border, where a 150-mile stretch of BR 364 remains incomplete. Its completion has been the talk of prime ministers and presidents in Tokyo and Washington. BR 364 would cut through virgin rain forest and connect Brazil to Peru, thereby opening up Amazonian timber for export to Japan via the Pacific. As projects go, it is not very big. By many estimates, it will cost only $300 million. In the early 1980s, the Inter-American Development Bank (IADB) was slated to finance the road construction. But with the international outcry over the murder of union activist Chico Mendes in 1988, the road became a hot potato and the bank has had to drop it. The bank's bureaucrats feared reduced funding from the Congress should they proceed.

Unfortunately for the environmentalists, the IADB is not the only game in town. The Brazilians are now flirting with Japanese financiers, who not only appreciate the commercial potential of the road, but also are less vulnerable to public opinion. In June 1988, the *Wall Street Journal* reported that Brazil is seeking $5.5 billion in official Japanese loans—so $300 million for BR 364 is peanuts.[14] Not surprisingly, there are rumors that the IADB is reconsidering the loan, and the joke among Brazilians is that the bulldozers will roll down the highway now that the United Nations Conference on the Environment and Development, RIO '92, is over and done with.

If the IADB loan does go through, then the game of playing one financier off another will have succeeded.[15] Now the Brazilians can play the same game regarding the legislation of equal protection. Say only Japan recognizes Brazilian property rights over genetic information. Exclusive Japanese co-ownership of genesteads will mean that Japan can create a stranglehold on newly patented pharmaceutical, chemical, and agricultural products that draw on complex natural genetic information.

The First Simultaneous Theft
of Artificial and Natural Information

The first simultaneous theft was probably that of *Chinchona*, the tree genus whose bark renders quinine.* The discovery of its therapeutic values was a closely guarded secret by the Quechua of the Andes. They must have felt vengeance satisfied as their Spanish plunderers died of malaria, a disease they knew how to cure. Such satisfaction ended in the seventeenth century when the Peruvian viceroy's wife, the countess of Chinchon, fell ill with the disease. Given the viceroy's wealth and power, it is little wonder that the secret GCF of chinchona would be leaked. Once the countess was cured, the plant took on her name. As malaria spread among Europeans, so did the demand for chinchon, which became the basis for quinine tonic water. The demand for tonic was literally stripping the world's supply of chinchon bark.

In 1859, a young Englishman, Clements Markham, was commissioned by the first secretary of state for India to bring out of the Andean region enough chinchona seeds for cultivation in British India. After much travail, this Markham did with a load of 100,000 seeds floating downriver to port in Guayaquil, Ecuador. The town mayor, realizing what this theft would mean to the local economy, ordered his arrest; Markham rerouted his exit through Peru, where he was able to bribe the minister of finance.

*Anthony Smith, *Explorers of the Amazon* (London: Penguin, 1990).

As countries like Brazil succeed in persuading just one country like Japan to respect equal protection, industries in all the other northern countries will be locked out of future opportunities that utilize the natural genetic information endemic to that southern country. No doubt realizing this, the other northern countries will come around. Ideally, the powerful lobbies in Washington, D.C., and their Eurocratic counterparts in Brussels will do an about-face and agitate for privatization for fear of Japanese dominance in their most profitable industries.

For the strategy to work, there must still be effective policing of virgin habitats against surreptitious collection. This will not be easy, inasmuch as information theft by Northerners is a tradition that spans centuries. The only solution is to deny access to virgin habitats to any country that refuses equal protection. Although this may be an ugly diplomacy, the South can resort to an eclectic rhetoric to justify the exclusion. The rhetoric could mix national security and biodiversity with the rights of indigenous peoples.

9

Finance

The answer pre-exists. It is the question that must be discovered.

Jonas Salk

Once nonpaying users are excluded from the genetic-information commons, several obvious questions emerge regarding contracts and finance. First, how will the genesteaders be remunerated in the interim between staking out the genestead and receiving the first royalty payments? Second, how will the countries of the South guarantee long-term ownership over their patrimony? Third, why will privatization be in the interest of the landed class as well as the landless class? Fourth, how will financiers of genesteads monitor their investment? And fifth, who will finance the gargantuan database needed to identify the claimants? Each of these questions is complex and, upon scrutiny, invites simpler and more specific questions.

How Will Genesteaders Be Remunerated in the Interim Between Staking out the Genestead and the First Royalty Payments?

It will take time and money to establish one's claim to genetic information. Recall from Chapter 7 that a representative from each group must be photographed according to protocols regarding, say, the buds, flowers, and seeds. Inasmuch as there are hundreds of species per hectare and the genestead is 160 hectares, we are potentially talking about thousands of photographs per genestead. Moreover, preserving these species is not just letting them be. Genesteaders must thwart squatters, remove nonindigenous species, and reduce edge effects. There are two possible sources of income: direct payments from financiers and revenues from extractive services. Both require a bit of explanation.

With co-ownership, investors, probably Northerners, will buy foreign debt at a discount and exchange it for local currency. There is a growing literature on just how such deals work. They are known as debt-for-nature swaps.[1] The swap enables a multiplier in local currency for the hard currency used. So the question of remuneration really becomes one of supply and demand conditions for discounted debt and for labor. For ease of calculation, let us assume the discount rate is 50 percent. In Brazil, the minimum annual wage is about $1,000. Again, for ease of calculation, assume that the cost of the camera equipment, tools for habitat management, and simple shelter are another $3,000 per year. Therefore, the genestead will cost $4,000 per year, which can be had through a debt-for-nature swap for $2,000. This is not much money; the financiers need not be the large industries. For example, most secondary schools in the North could find $2,000 in their budgets and justify the expenditure as part of the curriculum. The genesteader and the northern financier would be co-owners in the residual claim to the genetic information—that is, 7.5 percent each.[2]

Mainstream economists may see this as a share contract and object. They will object to it because share tenancy is inefficient *given* neoclassical assumptions.[3] Neoclassical analysis holds that tenants will reduce their wage labor by the percentage of the share when their take just equals that of alternative work. All this can be demonstrated with graphs and triangles, but let me explain in words. Suppose the genestead is owned 50–50 by the genesteader and the financier. If one day of labor-intensive genestead improvement is worth $10, the genesteader will improve the genestead only when he cannot get work outside the genestead that pays better than $5. Any wage greater than $5 is worth more to him than his cut of the improvement in the genestead (50 percent × $10 = $5). But inasmuch as a day's labor improving the genestead is in fact worth $10 and not $5, labor is not allocated to where it is most valued. There is an efficiency loss.

Because of this $5 loss of value, the mainstream economist would recommend against share tenancy. However, such a recommendation would ignore the relative efficiency of the alternatives. They are basically two:

1. The Northerner could buy the genestead outright and pay a wage to the genesteader.

Debt-for-Nature Swaps, Keynesianism versus Monetarism

Most of the world leaders embrace some sort of economic philosophy that derives from monetarism. Monetarism holds that an increase in the money supply only increases inflation and that inflation is a tax. In other words, stimulating demand through printing money does not result in more output, just more taxes.

With that said, how would a monetarist view the debt-for-nature swap? Because the debt-for-nature swaps result in more currency being printed in local markets to buy the land titles for the nature reserves, the end result of increasing the money supply is an inflation. Inasmuch as inflation is a tax, the debt-for-nature swaps are in essence a tax to recoup for the bad loans made in the past. These loans were often negotiated by dictators and for dictators. This logic leads to two questions: Should consumers in the South be taxed for debt-for-nature swaps? Should they help bail out the banks that financed the dictators? Such reasoning is not left wing; it derives from the economic philosophy of the central banks of the OECD: monetarism.

I posed these questions to the panelists at the RIO'92 Preparatory Congress on Comparative Environmental Law held in Rio de Janeiro in October 1991. Although all of the panelists endorsed debt-for-nature swaps, not one chose to answer either question. I asked the question not because I am against debt-for-nature swaps but because I am for them! The questions will occur to people who are adamantly against debt-for-nature swaps and who will use monetarist logic.

Anyone sympathetic to the debt-for-nature swaps should know that the monetarist conclusion is correct only when monetarist assumptions hold. Those assumptions are full employment and a flexible price mechanism. In the 1930s, British economist John Maynard Keynes challenged those assumptions for a depressed economy. Keynes argued that when unemployment is widespread and prices are rigid in the downward direction, expanding the money supply increases production. Today the Keynesian conditions hold in much of the South.

The monetarists will retort: you must be joking! Latin American governments routinely stimulate demand by printing money, and the result is rampant inflation. The paradox of widespread unemployment and inflation lies in the dual structure of most Latin economies. The cash economy is urbanized and closer to the monetarist conditions. The rural economy is outside the circulation of money and, therefore, closer to the Keynesian conditions. A debt-for-nature swap in which the money flows to the un- and underemployed in the rural economy will have the Keynesian effect—that is, increased production. A debt-for-nature swap in which the money flows to the urban centers will result in the monetarist effect—a consumer tax in the form of inflation.

2. The genesteader could pay a rent to the Northerner and claim all the royalties.

The problem with the first is that the transaction costs of policing labor would probably be greater than the efficiency loss of share tenancy (is the genesteader removing nonindigenous species, taking the photographs, and so on?). The problem with the second is that the income from the royalties is not a continuous stream; it may be 10 years before the renter sees a royalty—there would be no market for rent now. In poor countries, share tenancy is the optimal solution in the short run. However, mainstream economists are right in the long run. The optimal solution is ownership by those managing the genestead (getting all $10 for the $10 improvement). This brings me to the second question.

How Will the Countries of the South Guarantee Long-Run Ownership over Their Patrimony?

The answer: attenuation of the contract—that is, legal constraints on foreign ownership. Attenuation can actually enhance efficiency and is one of the many counterintuitive insights of property-rights analysis. In this case, constraints on foreign ownership should be designed so that foreigners can invest in the genestead and retain half the royalty rights for a limited tenure. In light of fairness and simplicity, the terms should be the same as those enjoyed by artificial information. For example, in the case of pharmaceuticals, that term would be 17 years. So if something patentable comes to fruition in those 17 years, then the investor will have patent rights for another 17 years from the date of patent. After the initial 17 years, his lease will have expired and he will have lost claim to any genetic information not under patent. Similar rights will apply to analogous forms of intellectual property— that is to say, copyrights, trademarks, and trade secrets.

Once the lease expires, the genesteader will have the option to buy out the foreign interests. Once they are bought, the genesteader becomes the full owner of future intellectual-property rights. Such a scheme would eliminate the efficiency loss of share tenancy. But there is a snag. To the extent that the genesteader can eventually become the full owner, what incentive will the investor have? Recall that he will have a 50–50 royalty claim to the genestead for 17 years. To sweeten the deal, the investor should also be given compensation for the 17 years over which he has supported the genestead.

The Best Parcels Go First

A basic principle of finance is to invest where the return is highest for any given level of risk. Accordingly, the hectares to be preserved first will be those most likely to yield genetically coded functions found nowhere else. Three criteria emerge for investment in any particular hectare: (1) the richness of the hectare in biodiversity (as measured by the number of species), (2) the harshness of the environment, and (3) the uniqueness of the biodiversity (as measured by the branching point with other species in the tree of life). Why these criteria?

1. The richest hectares are found in the Amazon and are usually those that survived the last ice age. Although the ice never descended into the tropics, the cold did. From 2.5 million to 10,000 years ago, much of the Amazon basin turned into savanna. That which did not later served as a refugia to replant the Amazon. Sixteen patches of rain forest have been identified as unusually rich as measured by the number of species.

2. Evolution explains why the genetic information from some habitats is bit for bit more useful than that of others. In a highly competitive environment like a rain forest, plants evolve complex biomolecules as a defense mechanism. Many of these mechanisms are now functional for medicine.

3. As an investor, one would also want to know how many commoners will claim royalties for the same bit of commercialized genetic information. Geology becomes the best indicator for estimating the size of the commons. For example, any 1 hectare of rain forest in Queensland, Australia, probably contains far fewer species than 1 hectare of the Amazons. Nevertheless, the uniqueness of the genetic information in the Australian species is probably far greater. The reason is geology: 150 million years ago, a big chunk of the original land mass (Pangea) split. That chunk (Gondwanaland) drifted northward, breaking its landbridge with South America about 50 to 60 million years ago to become Australia. Because of its geological history, Australia is likely to have more unique genetic information per species than that found on the other continents. The same is true of Madagascar.

In sum, the implication of natural history for finance is that the 16 patches of refugia in the Amazons and the small remaining virgin habitats in Madagascar and Australia will be the first bought.

So at year 17, he will receive the present value of the annuity compounded for 17 years. These payments were estimated at $2,000 per year. Assuming an average real interest rate of 3 percent, the calculation becomes $2,000 $(1.03)+2,000(1.03)(1.03) \ldots 2,000(1.03)^{17}$, or approximately $45,000 in uninflated dollars.

Because the value of the genestead in year 17 is determined by a formula and not by the market, many free marketeers will object. Why constrain the market? The answer is again the efficiency of attenuation. Foreign investors know best which properties have the greatest value; with asymmetric information regarding habitat value, the land values will be bid so high for the most biodiverse parcels that the genesteaders may never be able to afford to buy out the Northerner, and the inefficiency of share tenancy will never be eliminated. A formula gives the financier both a fair market return on the capital invested, 3 percent per year, and a 17-year window for windfall profits in royalties.

Suppose the genesteader cannot finance the $45,000 in year 17. Suppose nothing became commercialized and there are no royalties. In this event, the genesteader can ask the foreign investor to continue the share tenancy. If the foreign investor declines, then the foreign share goes back onto the market. If there are no buyers, then the genesteader will have to move out and let some other entrepreneur give the genestead a go; he should not be able to stay on without a stream of income because the risk of clandestine deforestation is too great—the costs of policing are too high.

Hopefully, international financing will not be necessary. A whole literature has sprung up on extractive reserves as I have been writing this book. The evidence seems overwhelming: the "sustainable" use of the rain forest for collection of edible fruits, nuts, rubber, and controlled hunting is greater than the use for cattle or timber. And these are just the traditional sustainable uses; there are potentially many more.[4] Perhaps these extractive reserves can pay the $4,000 per year minimum needed. If they can, then why has there been this deforestation and mass-extinction crisis in the first place? The answer lies in the question of landownership.

Why Is Privatization in the Interest of the Landed Class as well as the Landless Class?

Mutual benefits can be illustrated through the case of rubber production. The tappers have it tough. They eke out an existence by traveling days on end to tap scattered rubber trees, and tapping is in and of itself a nasty and

For the Man Who Has Everything . . . Give the Gift of Survival: A Beetle for Christmas

The attachment of one's name to a public good is often a tremendous incentive. Scientists have principles and theorems named after themselves; politicians have streets and cities, but what has the average environmentalist?

One possibility is a newly discovered species. One can easily strip the asset "species" into many attributes beyond functional value in agriculture, chemicals, and pharmaceuticals. The attribute of interest here is status.

The U.S. academician who may find this idea far-fetched need look no further than his ivory tower. The tower itself probably bears the name of some alumnus. Indeed, universities are notorious for selling status to the highest bidder. For example, in the 1920s tobacco tycoon James B. Duke wanted Princeton University to be the beneficiary of his fortune, provided it changed its name to—you guessed it—Duke. Princeton, which already had a reputation (and wealthy donors), declined. A poor normal college in North Carolina (Trinity College) accepted. Trinity became Duke University.

Fifty years later, in the construction of a student center, Duke University delineated the attributes of status to the nth degree. Why sell status to just one benefactor? What about the other alumni? One thousand middle-class alumni at $1,000 each will bring the same revenues as the $1 million donation. The principal benefactor would get his name over the doorway, and the middle-class donors would get plaques cemented into the wall along the footpath. Contributions were maximized even if architectural aesthetics were not.

There is no reason why ego mingled with altruism cannot also be applied to taxonomy. The first purchase of status through species classification was by film producer Steven Spielberg, who donated $25,000 to the Dinosaur Society. In return, "the society renamed the oldest known ankylosaur *Jurassosaurus nedegoapeferkimorum*. Part of the second word is an acronym for the surnames of the film's cast."*

There are other opportunities. Many avid bird-watchers are multimillionaires and may pay in excess of six figures to have a bird species or subspecies named after them. Although rare, newly discovered bird species are still being discovered. And for the yuppie on a tighter budget, there are the beetles. By some estimates, 30 million. If the profit of naming one beetle species after a benefactor is just $1,000, then one has generated $30 billion for conservation when all the beetles are named.

*Richard Corleis, "Behind the Magic of *Jurassic Park*," *Time*, 26 April 1993, 55–56.

dirty job. Their income is always uncertain, inasmuch as the market price of rubber fluctuates. Traditionally, the tappers have had a share contract: the landlord takes a percentage of the rubber tapped. But out in the bush, the landlord does not know how much each rubber tapper taps. Monitoring costs are high. Therefore, a feudal system has emerged in which the landlord minimizes the risk of theft: he exacts the death sentence for anyone even suspected of cheating. But the Middle Ages are over; the rubber tappers have discovered liberation theology and Marxism. Through unionization, they are demanding human rights. And through the international media, they now have a voice in governmental decision making. Times are changing.

Not to disparage the tappers, but property-rights analysis would indicate that once they believe the death penalty is no longer *in vigeur*, more will cheat. More cheating means that the monitoring costs of landlords will increase. The landlords probably reason that after rubber tappers secure human rights, land reform is next. Any legal challenge to the landlord's titles will, in the language of property-rights analysis, increase the future negotiation costs as the landlord loses a sense of certainty over his future ability

---

What Would Chico Think?

Chico Mendes was a union leader of rubber tappers in the Brazilian rain forest.* His murder in 1988 captured world attention: it has also been a windfall for both union membership and the rain-forest movement. The murder raised the transaction costs of landowners to delineate their property through deforestation.

What would Chico think of privatization as a conservation policy? He probably would not be impressed. Chico embraced an eclectic philosophy that mixed Marxism with a Gandhi-like pacifism. When environmentalists demonstrated that they could advance his cause, he embraced environmentalism too! Through the assistance of environmental activists, Chico springboarded himself and his cause to international fame. However, Chico had no time for property rights—in strict Marxist fashion, he believed in none. It is ironic that property-rights analysis can also explain the self-interest of such a position: deny the landowners any property right de jure, and the rubber tappers will enjoy them de facto.

*See Alex Houmatoff, *The World Is Burning: Murder in the Rainforest* (New York: Avon Books, 1990), or Andrew Revkin, *The Burning Season: The Murder of Chico Mendes and the Fight for the Amazon Rainforest* (Boston: Houghton Mifflin, 1990).

to consume and obtain income from his alleged property—the rain forest. Expropriation is a real worry. So what does the landlord do? He deforests. It is much easier to monitor a herd of cattle than it is to monitor the rubber tapped in this post–Chico Mendes era. And once the rain forest is deforested, the very attributes of the land that the tappers want will quite literally go up in smoke; what remains are the attributes that the landowner can easily assert control over—the attributes of pasture. Property-rights analysis explains the paradox of a rain forest worth more as an extractive reserve than as a pasture but nevertheless deforested.

Will both the landowners and the landless embrace privatization as a conservation policy? Inasmuch as the northern financing of $4,000 per year for 160 hectares would be forthcoming and, when combined with expected royalties and the value of extracted products, would yield an income per hectare many times greater than that obtained through cattle or timber, selfish interest implies that both landowner and rubber tapper would have a strong incentive to come to terms and reconcile.

How Will Foreign Financiers of Genesteads Monitor Their Investment?

In terms of creditworthiness, the South has next to none. Northern banks were burned badly by the debt crisis of the 1980s. Financiers will quite naturally be suspicious that their genesteads will become nothing more than paper parks. Indeed, no policy can rest on trust. The financiers must be able to monitor the genestead at relatively low costs. Fortunately, satellite images can fairly cheaply detect the most flagrant abuses of habitat destruction by fire.[5] However, selective destruction cannot be picked up by satellite. A satellite cannot discern things like poaching, mercury contamination of rivers, and selective logging. On the ground, random sampling will also be required. To assist inspection, genesteaders must be persuaded that deforestation in neighboring genesteads threatens their genestead and that it is therefore in their own best interest to report such deforestation.

Southerners may recoil at the notion of satellites spying on them from above or neighbors denouncing neighbors whose only crime is cheating a *foreigner*. Here, political leaders in the South should take the high moral ground and explain three simple facts to the public:

1. Without the satellite check, there would be no transfer of money to genesteaders.

A Thousand Points of Light

Part of the rhetoric leading up to the American presidential campaign of 1988 was candidate George Bush's vision of "a thousand points of light." No one (including Bush!) could state exactly what these lights were. But every one suspected it was American sop for hope and charity. Ironically, as the then-presidential hopeful palavered on, Brazil's Institute for Space Research (INPE) counted some 6,800 points of light on just one day. The images were taken from satellites hovering over the Brazilian interior. These lights were no metaphor for hope and charity. They were literally lights—points of fire—strung across vast stretches of the Trans-Amazonic highway. The lights meant greenhouse gas, genocide of Amerindian peoples, and the mass extinction of species.

2. Those genesteaders and inspectors who defraud the system are imposing a cost on every honest genesteader and inspector; to denounce them is to protect the honest.
3. Genesteaders whose neighbors have deforested will suffer increased costs due to worsening edge effects.

Because monitoring the genestead for habitat destruction is essential for finance, care must be taken in designing a system that does not turn the genesteaders into de facto police and institutionalize frontier justice. The genesteaders are not police. Instead, they should be viewed as communication links. Therefore, the structure of the policy should ensure a cascade of checks and balances. One possibility is a monthly transmittal stating whether or not the genestead is being encroached and whether or not the neighboring properties are being encroached. This information could be sent simultaneously to four bodies: the financier, a human-rights group like Amnesty International, the squatters, and the police. If the police do not act in a timely and just fashion, then the deforestation is not the fault of the genesteader and the genesteader should not suffer for it. Instead, he should be relocated. In the same vein, the financier who co-owned the genestead should not suffer. The money already invested should be reimbursed. This money will come from the same source that funds the gargantuan database. So the answer to How will the reimbursements be financed? is also the answer to the fifth and last complex question.

Who Will Finance the Gargantuan Database?

To answer this question, one must return to the baseline ethic: *Those who benefit, pay the costs associated with the benefit.* Just as the North benefits from the genetic information stored in the South and should pay the opportunity costs of such storage in the form of royalties, Southerners who benefit from the gargantuan database should also pay for the associated costs of its maintenance. The beneficiaries are those individuals who collect royalties on the genetic information.

The way the costs of the gargantuan database will be apportioned is similar to the way the royalties will be apportioned. To understand the apportionment, the highly technical language of property-rights analysis becomes useful. Returning to Harold Demsetz's seminal 1967 article cited in the boxes on pages 25 and 26, Demsetz writes: "An increase in the number of owners is an increase in the communality of property and leads, generally, to an increase in the cost of internalizing."[6] In the case of genetic information, "internalizing" refers to the cost of distributing the royalties to all the commoners of the same piece of genetic information and the fixed cost of the gargantuan database. To apportion these costs, an equitable and efficient rule must be designed. It must also be simple, inasmuch as simplicity is the key to persuasion. One possibility is the formula xR/n, where x is the number of organisms on the genestead that have the commercialized GCF, R is the royalties collected per year, and n is the total number of organisms of the taxon at which the genetic information is distributed: that is, $n = \Sigma x$. Because of the complications of counting many organisms distributed over many genesteads, it seems highly plausible that the variable costs (VC) of delineating the commons will rise precipitously with the number of landowners and/or the number of organisms that exhibit the GCF, just as Demsetz would suggest. Imagine distributing royalties for an alkaloid common to several species found all over the Amazon rain forest! To explain what should be done in the case of large x or n, I will now resort to the dreaded graphs of textbook economics (Figure 9.1).

Suppose ecologists estimate that there are n_1 organisms and that the cost of ascertaining their distribution over genesteads is R_1. However, if the total royalties ever derived from the GCF is only R_0, it would be economically wasteful to bother with the count, since the royalties would be less than the expected cost of the count R_1, that is, $R < R_1$. The transaction cost is greater than the benefit. However, future sales have an element of uncertainty. Over the lifetime of the patent, the product could boom (again, think of AZT), and at some future time, the royalties could cover the cost of the biological

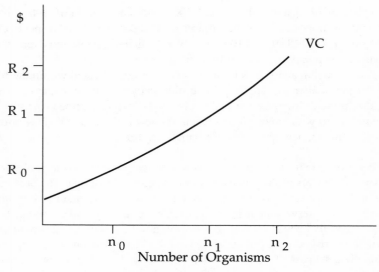

FIGURE 9.1. Royalties among commoners.

count. As soon as $R > R_1$, the count should begin with royalties remitting to the landowners on the basis of their share in the population pie, x/n. Until R_1 obtains, all royalties should be placed in an escrow account.

If over the useful life of the patent, R_1 never obtains, and the royalties only accrue to R_0, what should be done with R_0? On the basis of both efficiency and equity, R_0 should be placed in the sinking fund to reimburse the northern financiers who have suffered losses through clandestine deforestation. Hopefully, this will be a small fraction of the total money collected. The rest of it should be used to diminish the fixed costs of the gargantuan database. If all those fixed costs are eliminated through many cases of R_0, then the surplus money should be used to finance research pertinent to the habitats of the commercially valuable organisms. The logic here is that the industrial users will still be paying for the marginal cost of the genetic information as dictated by the baseline ethic, *those who benefit, pay the cost associated with the benefit.*

The same logic extends to how the variable costs of the count should be borne when $R > R_1$. Inasmuch as the genesteaders are receiving royalties proportional to their share of the genetic commons, they should also be bearing the variable costs proportional to their share of the genetic commons (xVC/n). So if Juan has only one plant and it was that one plant that was photographed, and Dolores has 1,000 plants, and Camilo, 10,000 plants, then the royalties as well as the variable costs should be shared respectively 1/11,001, 1,000/11,001, and 10,000/11,001. With the variable and fixed costs covered, the policy becomes self-financing.

The preceding analysis is by no means complete. Nested within the big picture are smaller ones that require resolution by a variety of experts across the disciplines. Moreover, the analysis is not static. It is not frozen in today's technology or with today's information. As new technology and new information emerge, new questions of finance will follow.

As mentioned in the introduction, it would be tempting to end the book here. Several reviewers of the manuscript have suggested just that. The problem is that there exists an overriding issue for this and any other policy that attempts to preserve genetic information. That issue is global warming. The case can easily be made that global warming will sustain the current mass-extinction crisis even if privatization were implemented today and enjoyed immediate success. To address what should be done is not easy. Global warming not only is a complex scientific phenomenon, but has become a complex political phenomenon.

10

Final Payments:
Greenhouse-Gas Abatement

A man hears what he wants to hear and disregards the rest.
Paul Simon and Art Garfunkel, "The Boxer"

There is a psychological phenomenon known as cognitive dissonance. It is a mental process by which one chooses only those facts that confirm prior beliefs. In other words, one filters the evidence and discards that which would force a revision of prior beliefs. Cognitive dissonance would explain why, say, the French Communist party is still as committed as ever even after the failed Soviet coup of August 1991. The party members simply discard the evidence and deny the rational inference—communism fails as a political system. Economists are now suffering the same syndrome when it comes to greenhouse gas.[1] They are discarding the evidence of global warming because it may force a revision of their prior beliefs. The result has been an almost total silence. This silence is not because the greenhouse effect is something new and exotic; it has been known for at least a century, heavily discussed during the past 30 years, and elaborately modeled over the past 10 years. The silence is because of cognitive dissonance.

Cognitive dissonance among economists creates an obstacle that climatologists may not yet fully appreciate. For example, a leading climatologist, Jeremy Leggett, writes "An understanding of the remarkable consensus among the world's climate scientists, and consideration of the impacts of the rates of global warming they predict, should surely lead any rational person to conclude that a concerted international response to global warming can no longer be delayed." This consensus came from "more than 300 of the world's most eminent climate scientists—in governmental service as well as in universities. According to Dr. John Houghton, the chair of the Intergovernmental Panel on Climate Change (IPCC) scientists group, the IPCC Report involved every serious researcher in the world in the field of climate-modelling, and 'less than ten' would disagree with its conclusions."[2]

The trouble with Leggett's line of reasoning is that it depends on persuading "any rational person." No one is truly rational; we all have a tendency to engage in cognitive dissonance and denial. Economists are no different.[3] Even champions of the prototypal "rational economic man" can be quite irrational when it comes to greenhouse gas. For example, Gary Becker, the 1992 Nobel Memorial laureate in economics, lambasts climatologists in a *Business Week* article entitled "The Hot Air Inflating the Greenhouse Effect." Becker warns sternly, "The extreme proposals to cut back carbon emissions rely on forecasts of global warming that are highly questionable. It's just one illustration of the misleading claims about many environmental risks."[4] And he is not alone among economists. William Nordhaus is even more belligerent. When it comes to doing something about greenhouse gas, Nordhaus advises otherwise and claims that "the policy cart has been careering far in front of the scientific horse."[5]

Climatologists had better dig in their heels. All indications indicate they are in for a knock-down, drag-out fight.[6] If global warming is indeed happening, then the economists have got it all wrong—they have failed to incorporate *the* causality that will drive resource allocation in the twenty-first century. Public acceptance of the climatologists' case will ultimately mean public rejection of economic theory and perhaps even ridicule of its practitioners. So it is in the selfish interests of economists to pooh-pooh the threat of global warming. For these reasons, I conjecture that if one took a poll of the world's 300 most eminent economists—in governmental service as well as in universities (measured by publications in ranked journals)— 290 would claim that neoclassical economics is the only scientific economics and, among these, most would probably agree with Becker and Nordhaus. Which consensus do we listen to? The climatologists or the economists? Perhaps the 290 climatologists are flat wrong and the 10 in the minority correct,[7] or perhaps my hypothetical 290 economists are wrong and the 10 in the minority are correct.

The philosopher Erich Fromm had the answer. He once remarked that "one can be a minority of one and still be correct." Science works that way: it is both undemocratic and fair. A simple consensus will not do. Theories must be logically consistent, and evidence must be presented in a way that

Economists on the Warpath

Theoretical economists seldom go public. Something really has to get their hackles up to stoop so low. That something today is greenhouse gas. In *The Economist*, William Nordhaus, professor at Yale and former member of President Carter's Council of Economic Advisers, speaks out with words that are either so surreal or so vituperative that I have chosen a dozen choice ones to illustrate the angry spirit in which they were written.*

1. "Even a distinguished international panel of scientists, who should know better, calls for a 60% cut in these emissions."
2. "Much is conjectural in such (climate) forecasting, and few climate modelers expect to improve their forecasts dramatically in the near future."
3. "Climate warming will probably be a boon to Alaska, which is America's least productive state in GNP per square mile."
4. "Studies suggest that greenhouse warming will reduce yields in many crops, but the associated fertilization of higher CO_2 will probably offset any climatic harm over the next century."
5. "In recreation, snow skiing will be hurt, but water skiing will benefit."
6. "Climate change is likely to have less effect than the economic reunification of Germany this summer."
7. "[E]very time I read of a new deadly tropical virus, I wonder whether humanity could do with a little less biodiversity."
8. "[H]aving school children plant billions of trees would probably do more for civic virtues than for the climate."
9. "There are simply no substitutes for many of today's fossil fuels."
10. "The best investment today may be learning about climate change rather than preventing it."
11. "A bolder step would be for brave political leaders to launch an international Manhattan Project to develop safe nuclear power."
12. "Like other religions, the environmental movement needs a catechism of homilies."

*William D. Nordhaus, "Greenhouse Economics: Count Before You Leap," *The Economist*, 7 July 1990, 21–24.

invites scrutiny and challenge. This book is not the place to debate whether the evidence submitted by climatologists is sufficient to accept the consensus opinion. However, with respect to logical consistency, this is very much the place to discuss the greenhouse effect and neoclassical economics. The criterion of logical consistency may be sufficient to dismiss the economists' protestations. Once those protestations are dismissed, we can then plug privatization as a conservation policy into various policy proposals on greenhouse gas abatement *should the climatological evidence be persuasive.*

A very basic argument for the greenhouse effect and against mainstream economics is one that seldom enters the public debate: the second law of thermodynamics, also known as the entropy law. The second law implies that there is an irreversible degradation of energy gradients from order to disorder. In the degradation of resources, amplification effects are common. The future becomes hostage to the present, and random events can take us up unforeseeable paths. Thermodynamicists call the present "boundary conditions" and the random events, "bifurcation points." In the degradation of energy gradients, order emerges as a means to speed up the process. The

---

From Dinosaurs to Humans to Cockroaches

The last time there was such an abrupt change in boundary conditions of the Earth's energy gradient was at the end of the Cretaceous period, 65 million years ago. In 1980, geologist Walter Alvarez offered a cataclysmic hypothesis to explain the change: the planet suffered an asteroid collision. Over the next 10 years, Alvarez and his followers gathered much physical evidence to support the hypothesis. In 1991, sediments were discovered on the ocean floor off the Yucatán Peninsula, that would indicate the collision of an asteroid. And by 1993, evidence of the gravitational field in the crater indicated a diameter of 300 kilometers. To create such a huge crater requires a catastrophe 10,000 times greater than that of the entire nuclear arsenal if it were detonated at the same time! According to the cataclysmic hypothesis, the dust kicked up blocked out the sunlight and caused a rapid cooling of the planet.*

As the dinosaurs exited Earth, they abandoned niches for mammalia—our ancestors, which were then no bigger than rodents. Sixty-five million years later, the descendants of these rodents, *we the chosen*, are setting new boundary conditions through a consumption and reproduction frenzy. The new boundary conditions will clear the stage for the radiation of present forms of life resistant to the harsh realities of the greenhouse world. One of the most resilient of these species is the cockroach—a truly remarkable survivor. In the history of life, the next geologic epoch may indeed be that of the cockroach.†

*R. A. Kerr, "Looks Like the Yucatan Holds a Killer Crater," *Science*, 15 November 1991, 943; Virgil L. Sharpton et al., "Chicxulub Multiring Impact Basin: Size and Other Characteristics Derived from Gravity Analysis," *Science*, 17 September 1993, 1564–1567.

†Jonathan Schell, *The Fate of the Earth* (New York: Knopf, 1982), chap. 1.

emergent order is called, quite appropriately, "dissipative structures." The paradigm integrates so nicely with the greenhouse effect that one of its popularizers, Jeremy Rifkin, has changed the title of his book *Entropy* to include, in the second edition, the subtitle *Into the Greenhouse World*.[8]

The second law implies that the planet is emerging with new boundary conditions to dissipate the available energy gradients. Seen in this light, man is a dissipative structure who is setting those boundary conditions through the deployment of technology. Given the possibility of amplification effects through technology, it is quite possible that, if we continue our business as usual, we may be clearing the way for new forms of life more adaptive than ourselves to a greenhouse world.

Whereas global warming is logically consistent with the second law, mainstream economic methodology makes no thermodynamic sense, and, amazingly, mainstream economists readily admit it! This was demonstrated in a 1988 session of the American Economic Association convention entitled "Economics and Entropy." After all the papers were presented, every speaker (including Nordhaus) agreed that the entropy law has value only as a metaphor. In other words, neoclassical theory would remain intact and resilient to any reduction. The panel members were unanimous, and I believe they were unanimously wrong.[9]

Economic theory makes no thermodynamic sense because economic decisions and their consequences are assumed to be reversible. Reversibility means that economics has no boundary conditions, no bifurcation points,[10] and no unforeseeable amplification effects. In short, economics has no history. This lack of history extends beyond the degradation of natural resources like the atmosphere and reaches to human behavior itself. Without history, there are no explanations as to why humans exhibit nonrational behavior.

Because the greenhouse effect is an entropic phenomenon with far-reaching implications for resource allocation and yet cannot be accommodated by the standard methodology, the greenhouse effect has become a threat to mainstream economists. The possibility of an entropic phenomenon becoming just such a threat to theory and theoreticians was foreseen in the 1920s when physicist Sir Arthur Eddington wrote:

> The law that entropy always increases—the second law of thermodynamics—holds, I think, the supreme position among the laws of Nature. If someone points out to you that your pet theory of the universe is in disagreement with Maxwell's equations—then so much the worse for Maxwell's equations. If it is found to be contradicted by observation—well, these experimentalists do bungle things sometimes. But if your theory is found to be against the

second law of thermodynamics I can give you no hope; there is nothing for it but to collapse in deepest humiliation.[11]

Given the logical consistency of the greenhouse effect vis-à-vis the second law and the logical inconsistency of economics,[12] it is only logical that those in power—that is to say, bifurcation points like Clinton, Major, and Kohl—listen to the climatologists and not to the economists. Even better, the world leaders should probably put economists in their place. That place is not to engage in rhetorical excess and decry climatologists as doomsayers. The place for economists is to flesh out alternative policy regimes for greenhouse gas abatement *should the climatological evidence be persuasive.* In doing so, economists should cease and desist their unreasonable demand for certainty in the prediction of global warming. After all, uncertainty is why we have a multibillion-dollar insurance industry and a dozen academic economic journals dedicated to it. The only legitimate role for the economist is to illustrate "insurance" policies against greenhouse scenarios.

A basic principle of insurance is to identify all possible scenarios and assess the probabilities of each scenario given the best available knowledge. One begins with the very worst case scenario and then moves through the more benign ones. The very worst case is, of course, a runaway greenhouse effect like that which occurred on Venus. Possible but how likely? Sufficiently probable to insure against? A good analogy can probably be made with a nuclear exchange between superpowers. Through all-out war, we could conceivably bring on the winter and the end of humankind. Or it could cause a localized holocaust like Hiroshima and Nagasaki. In any event, the superpowers spend billions of dollars in insurance to prevent all-out war.

Just slightly more benign than a runaway greenhouse effect are the multitude of scenarios involving a rise in the sea level. The one that captures my imagination is the collapse of the West Antarctic ice sheet.[13] The ice sheet is several kilometers thick, and its melting would result in a 6-meter rise in sea levels. Possible but how likely? Sufficiently likely to insure against?

To know how much money to spend and how to spend it, one must first know how each scenario will eventuate. In other words, one must trace the deterministic pathways to the greenhouse world. Unfortunately, each pathway is complex and many are barely known. One of the most harrowing takes place in the deep and has all the makings of a horror film. On the bottom of the ocean troughs are lattice-like structures of methane and water known as methane hydrates. The solid hydrates create a shell up to 300 meters thick under which gas has collected for eons. It is estimated that there are qua-

The Many Roads to Hell

Because there are many roads to hell, there will be many premia to pay. A common feature of the roadside seems to be methane, a greenhouse gas that does not get its due.* Methane (CH_4) is perhaps 60 times more potent, molecule per molecule, than carbon dioxide (CO_2). Besides the solid methane hydrates in the ocean troughs, methane is also emitted in the 5 million square kilometers of bogs and marshes as well coal seams, melting permafrost, and the guts of some 250 billion termites. Some will see these sources as natural and ask: Is man to blame if Mother Nature passes gas? The answer is *yes*. As the planet heats up as a result of, say, the release of CO_2 from fossil-fuel consumption, the permafrost will melt faster and more methane will be released. More methane released means more global warming, and more global warming means more permafrost melting—an amplification effect.

Responsibility should likewise be assumed for the termites. When we deforest and create grasslands for cows (another source of methane), we are also expanding the niche for termites. Termites, like all insects, are capable of a population explosion, which subsequently means an explosion in the methane they release. And once methane is released, man is also responsible for its longer "lifespan" in the atmosphere! The reason is chemical competition: by burning fossil fuels, we emit carbon monoxide (CO), which competes with methane for atmospheric hydroxyl radical (OH·). It is OH· that cleanses the atmosphere of methane and other pollutants. Because OH· is now cleansing the atmosphere of CO, there is less OH· to cleanse the atmosphere of CH_4.

*Fred Pearce, "Methane: The Hidden Greenhouse Gas," *New Scientist*, 6 May 1986, 37–41.

drillion kilograms of carbon trapped in these structures, more than the known coal reserves of the world. Low temperatures and high pressures keep these shells intact and down on the ocean floor. But warming waters from above could penetrate the deep, create cracks in the shells, and send up plumes of methane. This is not pure speculation. Soviet scientists have observed plumes 500 meters long in the Sea of Okhotsk. They have even sensationalized their findings by suggesting that such plumes could explain the disappearance of ships in the otherwise peaceful waters of the Bermuda Triangle.

To insure against an event like fissures in the methane hydrate shells requires the collaboration of experts from various disciplines:

1. The climatologists would have to model what would be the greenhouse effect of releasing the equivalent of all the known reserves

of coal into the atmosphere (such as the collapse of the West Antarctic ice sheet).
2. The economists would have to put a dollar value on the damage done (for example, the submersion of Bangladesh).
3. The oceanographers and geologists would have to assess the probability of the shells cracking.
4. Perhaps most importantly, the politicians would have to communicate the premium calculus to the public. Anything less than the dollar value of the damage done multiplied by the probability of the methane being released is the minimum premium we should be willing to pay.

The methane hydrate shells are just one of the roads to hell. More exist: landfills of organic rubbish, the photochemical reactions of hot asphalt roads, fertilization of rice paddies, the burning of grasslands—and these are just the scenarios for methane.[14] Other potent greenhouse gases such as chlorofluorocarbons, nitrous oxide, and halons also have myriad sources. And of course, there is that old mainstay of the greenhouse effect, CO_2. It is easy to become pessimistic and despair that all roads lead to hell. And so it easy to engage in cognitive dissonance and denial. Courage is required on the part of the world leaders to confront the evidence and revise former beliefs in light of it. If a hell is in the offing, and 290 of 300 climatologists believe it likely, then world leaders must accept the bitter truth that the premia must be paid— and paid now. They must engage the scientific community to answer over and over again the same question for each scenario: How to pay the premia?

There will be no simple answer and no one answer. Any suggested measure to abate greenhouse gas must undergo intense scientific debate. Without such debate, a quick and dirty solution may even accelerate global warming! An example of such miscalculation concerns CO_2. Europeans are considering converting electricity-generating stations from coal to methane. Per watt of electricity generated, burning coal produces almost twice as much CO_2 as burning CH_4. It would seem sensible to convert from coal-burning plants to methane-burning plants and halve the greenhouse-gas emission. Right? Wrong! Or so says a British consulting firm, Earth Resources Research. Because CH_4 is so potent, the question of greenhouse-gas abatement really reduces to a question of entropy and dissipation. How much CH_4 is leaking out? The leaks begin with sea well-heads and high-pressure pumps and go through the distribution network right up to the site of end use. To measure the leaks is no simple task inasmuch as there is a strong incentive for natural-gas companies to under-report—they are winners if greenhouse policy means substitution of coal for methane. Earth Resources Research has done an ingenious break-even analysis. They calculate that if the leaks are greater than 2.8 percent, a percentage that seems optimistically small to anyone in the know, then there is no greenhouse-gas abatement in the conversion from coal to methane. However, if the leaks are 10 percent, as some suggest, then the conversion is yet another road to hell!

No doubt the Nordhauses and Beckers of the world will be unpersuaded by this line of reasoning. They will simply insist on more evidence of global warming and continue the cheap shots about the policy cart and the scientific horse. The most reasonable reply to the unreasonable requests for certainty is that made by Stephen H. Schneider of the National Center for Atmospheric Research: "Many critiques somehow understress the fact that the sword of uncertainty has two blades: that is, uncertainties in physical or biological processes which make it possible for the present generation of models to have overestimated future warming effects, are just as likely to have caused the models to have underestimated change."[15]

Lest I give the wrong impression, let me now say that not all economists who work on the greenhouse effect are doing battle with the climatologists. Many are doing exactly what economists should do: economics. The names that come to mind are Joshua Epstein, Raj Gupta, Michael Grubb, Terry Anderson, Donald Leal, Peter Hartley, Michael Porter, Jos Haynes, and Brian S. Fisher. These economists are explaining the various insurance policies that would abate greenhouse gas should the public be convinced that global warming is a contingency worth insuring against. A policy menu has emerged from this new and exciting literature.

Two of the most comprehensive policy reports are Michael Grubb's *The Greenhouse Effect: Negotiating Targets* and Joshua M. Esptein and Raj Gupta's *Controlling the Greenhouse Effect: Five Global Regimes Compared.* Both monographs cover much of the same ground and shed light on what economic theory has to offer in the way of solutions. Epstein and Gupta break down the possible policies into five alternatives:

1. A system of across-the-board equal percentage emission cuts
2. A system in which future emission reductions would be proportional to past emissions
3. A system of per capita emission targets
4. An emission tax system
5. An international market in emission permits.

Although the monographs are written for a specialist audience,[16] one does not have to be a specialist to anticipate the folly of the first three regimes

Did They Die in Vain? The Petroleum Wars

In 1989, Michael Grubb wrote: "Even in the U.S., no President has yet dared to introduce a petroleum tax remotely comparable with that in most industrial countries despite the obvious and almost undisputed strategic and environmental benefits which it would bring."* Within a year of publishing that observation, Grubb would probably have to revise his opinion. Not only do Americans not want a tax, they will even risk their lives to keep the price of petroleum down. President Bush made this perfectly clear as he prepared for Operation Desert Storm in July 1990. The pending war was "to keep our way of life." This rhetoric proved distasteful and by August gave way to worries about another Hitler and another Poland.

When it comes to risking lives for the sake of petroleum, the British are no better. The 1982 British–Argentine war over the Malvinas/Falkland Islands only makes sense in terms of the potential petroleum reserves in the territorial waters of the South Atlantic. For the pounds spent on the war, each of the 1,200 Falkland Islanders could retire to the United Kingdom a multimillionaire.

A greenhouse-gas accord will remind us of the lives wasted in the petroleum wars. They died not only for a mercenary reason, but for a miscalculated one at that! Under a permit system, the demand for petroleum will plummet, making the commodity extremely cheap and the permits extremely expensive.

*Michael Grubb, *The Greenhouse Effect: Negotiating Targets* (London: Royal Institute of International Affairs, 1989), p. 40.

considered. Across-the-board equal percentage emission cuts would be costly for fuel-efficient countries like Japan and relatively cheap for fuel-profligate countries like the United States. Such a policy would not float politically. It would fail on grounds of fairness, what economists call "equity." If the world chooses the second option, future reductions proportional to past emissions, then the dirty-rich countries will have to spend exorbitant amounts to reach the goal when those same dollars spent could abate more greenhouse gas if applied in the dirty-poor countries. Such a policy would fail on the criterion of efficiency. The third option, a system of per capita emission targets, also suffers a fairly obvious flaw. Energy-intensive countries would be plummeted into depression as they tried to meet worldwide per capita emission targets, while energy-unintensive countries would have no incentive to conserve until they reached the target.

The choices really boil down to just the fourth and fifth options. Choosing between them is difficult without some elaborate economic analysis. The fourth option, an emission tax system, is not nearly as straightforward as it seems. In theory, taxation would raise the price of fossil fuels, thereby reducing consumption and emissions; the tax revenues would be used to subsidize greenhouse-gas-abatement activities. Although seemingly simple, the institutional problems are complex.[17] First, in terms of consumption of fossil fuels, consumers are fairly unresponsive to the price of fossil fuels in the short run. Indeed, they are more likely to respond politically to the leaders who have imposed the tax than curtail their fossil-fuel consumption. Because

---

Beyond Energy Efficiency

Greenhouse-gas abatement goes well beyond energy efficiency. It is not just switching from fossil fuels to solar and wind power. It reaches right down to the clothes we wear and the food we eat. Final payments mean a massive alignment of incentives for seemingly benign activities. For example, if one wears a leather jacket and eats hamburgers, that's O.K. as long as one pays for the methane-rich flatulence of the cows (73 kg CH_4/year $\times$ 60 CO_2/CH_4 = carbon dioxide equivalent of 4,380 kg) that surrendered their skin and flesh. These revenues will then finance carbon up-take activities like reforestation (20 tons of carbon/hectare/year $\times$ 1016.06 kg/ton = 20,321.2 kg). Roughly for every five cows you eat, you would have to plant 1 hectare of trees.*

The cultural resistance to the final payments will be enormous in the meat-and rice-eating societies. Brazilians, for example, may love the idea of charging royalties for the genetic information of the Amazons, but hate the idea of paying for the greenhouse-gas contribution of their diet: heavy in both rice and beef. Politicians will have to explain that the royalties and the taxes are flip sides of the same ethic: *those who benefit, pay the costs associated with that benefit; and those who generate a cost, pay that cost.*

Besides efficiency, there is also an equity argument for these final payments. Why should vegetarians subsidize carnivores to the tune of global warming? For example, Hindus are morally opposed to the killing of animals that, like man, are highly sentient beings. And humanists may be morally opposed to Hindus who would forgo a human life to save a flatulating cow. Privatization of the atmosphere would end the unfair subsidization of human carnivores and sacred cows.

*Based on data from S. Brown, A. E. Lugo and J. Chapman, "Biomass of Tropical Tree Plantations and Its Implications for the Global Carbon Budget," *Canadian Journal of Forestry Research* 16 (1985): 390–94.

leaders live their political lives in this short run, they have no incentive to embrace a long-run solution like the tax. Besides this shortcoming, there is another fairly obvious flaw. An international carbon tax is highly regressive: it would hurt poor people in both the rich and the poor countries. The South will undoubtedly appreciate this and resist. But without the cooperation of the South, any system to abate greenhouse gas is futile. Australian economists Jos Haynes and Brian S. Fisher make this point very clear:

> Overcoming any within-country market failure problems is an economically rational course of actions with benefits in its own right, but it can only be a partial response to the perceived greenhouse problem. It would retard the rate of growth of greenhouse gas, not prevent growth. Multilateral action is therefore unavoidable if the problem is perceived to need a solution.[18]

With four policies down, the fifth and final option, an international permit system, seems the only chance for something both equitable and efficient. A permit system would begin by setting a target for atmospheric greenhouse gas. That targeted level of greenhouse gas would then be viewed as a commons.[19] The criterion of equity suggests that we should all have an equal share in it. How large would the commons be? Some climatologists believe it must be as small as 20 percent of the current emissions. With such a small greenhouse-gas commons, a market for permits would quickly bid up the price of permits and induce substitution into gas-abatement activities wherever possible. However, when energy-efficient technologies are either impossible or exorbitantly expensive, permits would be exchanged for money, and fossil fuels would be burned where they are most valued. Individuals who like consuming a lot of energy (the average Australian consumes 41,000 kilograms of carbon per day) would pay those who consume very little energy (the average Zairian, 31 kilograms of carbon per day).[20] Noncompliant countries would face stiff trade penalties, the revenues of which would be used to retire greenhouse-gas permits elsewhere.[21]

One drawback to the permit proposal is that the Zairians would have an incentive to plan large families in order to garner more permits. Grubb has carefully weighed these concerns and suggests that the rights be distributed on the basis of adult population. This makes sense, inasmuch as no country in the North or the South plans a generation in advance. There would virtually be no incentive to overpopulate. However, other issues of fairness will undoubtedly surface, issues like having to forgo coal-fueled industrialization or petroleum exploration. Grubb concedes that "the greenhouse problem is so large and complex that there will be no end to the fairness debate, for fairness on all counts is impossible. The only way out of the dilemma

---

The Greenhouse-Gas Equivalent of Herman Kahn's "Doomsday Machine"

The shrillest cries of unfairness will be from the fossil-fuel-rich countries. The producers will protest that they are bearing the brunt for any greenhouse accord. In effect, their protest will be over their right to continue polluting the fossil-fuel-poor countries!

Imagine that China, with its 300-year supply of coal deposits, demands to be paid not to pursue a coal-fueled industrialization. Such an argument has no ethical foundation. It is the moral equivalent of blackmail: "Pay us not to take us both into the greenhouse world." Imagine that one of the nuclear powers created a superatomic bomb, what the futurist Herman Kahn once called the "doomsday machine."* The machine is sufficiently powerful that, after detonation, nuclear winter ensues. "Pay us not to commit global suicide." What would be the response to this threat? Protests of unfairness over fossil-fuel reserves are in essence the same argument.

*Herman Kahn, *On Escalation: Metaphors and Scenarios* (London: Praeger, 1965), 227–28.

---

would seem to be the establishment of the simple and fundamental guiding principle of adult per capita rights to the atmospheric resource."[22]

Vis-à-vis the alternatives, a permit system for greenhouse gas seems to be both the most efficient and the most equitable policy. All the aforementioned economists agree. A permit system also dovetails with privatization as a conservation policy on both a theoretical and a practical level. On the theoretical level, a permit system is really privatization of the atmosphere, very much akin to the extension of property rights to genetic information. On the practical level, a permit system will require much of the same data regarding land use that is required by the gargantuan database. But these are not the reasons that an international permit system on greenhouse-gas emissions must be included in this book. The reason is property value. Global warming could easily become the leading agent in mass extinction and devalue the newly created intellectual-property rights over genetic information.[23]

To explain how global warming could easily become the leading agent in mass extinction, a little biology goes a long way. It is a well-known fact that species migrate very slowly with changing temperature. The faster the change in temperature, the fewer the number of species able to migrate. To make matters worse, there is often no land bridge to a cooler habitat. Much

Can New Petroleum Reserves Be Tapped?
The Case of Argentina

In 1991, President Carlos Menem of Argentina announced that he would privatize the state-run petroleum monopoly—Yacimentos Petrolíferos Fiscales.* The initiative would open up 140 areas of Argentina to exploration. Today Argentina is self-sufficient in petroleum and has an estimated 10-year reserve. The new plan hopes either to make Argentina an exporter of petroleum or at least to maintain its self-sufficiency beyond the 10-year horizon. In terms of the greenhouse effect, the privatization of Yacimentos Petrolíferos Fiscales will undoubtedly mean an increase in greenhouse gas. The question arises: Is privatization the problem and not the solution? To conclude yes would be to confuse the means with the ends. Privatization is an efficient means to harness petroleum, and it is also an efficient means to allocate the right to burn the petroleum. The end of the former is more greenhouse gas; the end of the latter is less greenhouse gas.

An international permit system that caps worldwide demand at 20 percent of current usage does not dictate how that 20 percent will be produced. For example, if the Argentines can produce petroleum more cheaply than can the British, then Argentine petroleum should replace British petroleum. This means that countries which cannot easily extract crude oil—the British in the North Sea and the Americans in the Arctic—will be the big losers. Again beware of vested interests: the BPs and Exxons will be gung ho for the privatization of things like Yacimentos Petrolíferos Fiscales, but utterly aghast at privatizing the atmosphere—the dichotomy is in their selfish interests and at the expense of everyone else.

*"Yacimentos Petrolíferos Fiscales," *Wall Street Journal*, 28 March 1991, B5.

---

of the remaining biodiversity remains on islands separated by agricultural land use. Metaphorically speaking, these species will be driven into a sea of human land use. Without a permit system on greenhouse gas, there is no effective way to diminish global warming; and without the privatization of genetic information, there is no incentive to rescue organisms even if global warming is diminished. The task will fall to government, often the same governments that have helped transport us into a greenhouse world. For these reasons, the privatization of greenhouse-gas emissions cannot be decoupled from the privatization of genetic information. Both are necessary, and neither is sufficient for an effective and equitable conservation policy.

11

Conclusion:
Ten Principles for Conserving
Genetic Information

The late Al Eichner, one of my professors in graduate school, edited a volume of essays entitled *Why Economics Is Not Yet a Science*. A summary of the essays appeared in *Nature*, and the summary itself triggered a backlash by neoclassical economists. In the same journal, Partha Dasgupta and Frank Hahn of the University of Cambridge chastised not just Eichner but *Nature* for printing such "nonsense"![1] With fear of similar reprisal, I now take Eichner's reasoning one step further: economics not only is "not yet a science," but is a false science. Like a false prophet, a false science can take you exactly where you do not want to go. It is false for basically three reasons:

1. It does not incorporate irreversibility.
2. It ignores the fact that human behavior is the product of evolution.
3. It suppresses all ethics other than the utilitarian.

By not incorporating, by ignoring, and by suppressing, economists have helped transport us into a greenhouse world, where genetic information is expunged and hundreds of millions starve.[2]

Nowhere is economics less a science than in the realm of biodiversity. Because the benefits of biodiversity defy quantification, economists simply omit them from their analysis, knowing full well that this omission protects their methodology.[3] They must think it will go unnoticed. But they are wrong. Biologists are not dumb. Biologists know that the omission is equivalent to counting the benefits as zero. So the maneuver backfires, and economics as a profession falls yet another notch in esteem.[4]

How can economics become a science? The answer is fairly obvious: economists must assimilate the findings of the natural, physical, and behavioral sciences.[5] Usually these findings are complex. Because the framework of economic theory cannot accommodate these complexities, the only alternative is to start with the findings themselves and employ those techniques from economics that work. In this book, I have attempted such a construction by reducing biodiversity into information theory and then deploying the economic analysis of property rights. What emerges is the implication that privatization can become a conservation policy.

Unfortunately, such a reduction is incomplete and fails to place the policy into the big picture. Seeing the policy in the big picture is probably the most effective rhetoric in promoting it. To do so requires a full reduction, right down to the ultimate causation of genetic information. Such a reduction can be had in nonequilibrium thermodynamics (NET). In this conclusion, I will reintroduce NET as the ultimate causation of genetic information. When seen as a small part of a big whole, privatization can challenge its alternatives on many fronts.

A basic tenet of NET is that order emerges through the dissipation of order. This means that nothing is sustainable in the long run. All life rides on a crest of energy and its dissipation. To facilitate dissipation, a staggering amount of order has slowly accumulated over geologic time. This order is embodied in organisms as DNA—genetic information. Through the rapid degradation of the environment in human time, this same information is now being expunged for a variety of land uses. In the language of NET, privatization is a means to channel the path of dissipation through time and take us where we want to go. Judging from the aspirations of people around the world, that place is consumer paradise. To get there, we must retain genetic information that has a high probability of future value.

The thesis of this book is that the best way to retain useful genetic information is to create property rights over it and, at the same time, constrain this newly created property. Orthodox economists will balk. They are taught to think that constraints reduce efficiency and that the reduction in efficiency translates into a reduction in human welfare. Property-rights analysis teaches us something very different. Constraints can actually enhance efficiency by reducing transaction costs. This is a very counterintuitive implication. It means that if we collectively exercise our right to surrender rights, each of us can be better off—indeed, much better off. The imposition of such constraints is called *attenuation* in the language of property-rights analysis.

It is in this vein that I and many other unorthodox economists are suggesting a sweeping creation and concomitant attenuation of rights. A permit to release greenhouse gas is the creation and attenuation of the right to pollute. A permit on childbirth is the creation and attenuation of the right to lever dissipation in the future through procreation now.[6] And privatization as a conservation policy is the creation and attenuation of intellectual-property rights over genes. The creation and attenuation of all these rights will put us on a path of dissipation through time that yields greater human welfare than the other paths. To put us on that path, government is very much required. Standards must be set and enforcement guaranteed. The costs associated with both the standards and the enforcement will be far less than those of an unconstrained economic system: a greenhouse world, mass extinction, and several billion more people jammed onto Earth.

The logic of this argument and any evidence for it will not be enough to persuade. Environmental policy, like economic policy, is the product of emotion and rhetoric. As such, policies will be adopted and implemented for reasons other than logic and evidence. Environmentalists seem to accept this and believe that the most effective rhetoric is a moralistic appeal. Here

The Selfish Interest of Environmentalists, Rich or Young

Fundamental to economic theory is the notion of diminishing marginal utility. All it means is that one's satisfaction diminishes with greater consumption. The fortunate few who can afford abundant private goods derive less and less pleasure from incremental consumption of these goods. In contrast, the consumption of public goods such as a pristine shoreline and clean air is beyond even a king's ransom. Quite naturally, the wealthy are closer to satiation in sundry houses, cars, and jewelry than they are for public goods such as a pristine shoreline and clean air. In terms of satisfaction, they stand to gain a large increment from environmental protection. Therefore, it is predictable from rational self-interest that the wealthy should be passionate about environmental protection.

The young, without property or income, are also enjoying some measure of environmental protection without having to foot the bill. Although not all environmentalists are wealthy or young, disproportionately more are wealthy or young than are working class. Inasmuch as the environmentalists are seeking more protection through the only means that seems forthcoming—greater government intervention—it will be the working classes who will end up paying for the bulk of that protection through the usual mechanism: taxation. Privatization is not only an efficient, but also an equitable, alternative to intervention and taxation.

NET cannot compete. NET is not a rhetorical enterprise; it is a science devoid of morals. But, nevertheless, as a science it does have a hidden advantage. It can put to rest much of the tired rhetoric of the environmental dialogue and, in the vacuum created, persuade by logic and evidence.

Environmental dialogue is tired because its most respectable advocates, the evolutionists, can no more calculate the value of biodiversity than can the economists. But whereas economists ignore the incalculable benefits in a dishonest effort to rescue their cost–benefit methodology, the evolutionists reject the methodology and look elsewhere for an effective rhetoric. Usually that rhetoric adopts some form of "green ethic."[7]

Lay converts have picked up the rhetoric and harangue the public. A fairly typical example is Queen Beatrix's 1988 Christmas message to the people of Holland: " The Earth is slowly dying, and the inconceivable—the end of life itself—is actually becoming conceivable. We human beings ourselves have become a threat to our planet."[8] From the scientific perspective of NET, we know that the Earth has always been dying. As dissipative structures, "we human beings ourselves" have always been a threat to the planet; only through destruction did we emerge as human beings. Second, the end of life is not only conceivable, but inevitable. Nothing is sustainable; ecosystems, like individuals, experience death and succession.[9] So open up the gifts and have a Merry Christmas—is this the NET message? No. The message is attenuation—choosing gifts for their impact on the future ability to give gifts.

This much NET advocates can concede to the queen: the concept of sustainability sounds noble and probably triggers deep-seated emotions that were once adaptive.[10] When voiced, the natural response is to applaud. However, when it comes down to acting on the rhetoric, an even more adaptive behavior prevails—selfishness. The advice offered here is cynical and may even sound depressing: let us rise above the biological urge to applaud the rhetoric, and let us be honest. There is no common heritage of mankind; all of us are primarily looking out for our own interests. Selfishness prevails, and biophilia is too weak to drive conservation. These have been the working assumptions of *Genes for Sale*.

From these assumptions, privatization emerges as a markedly different policy from the competition that calls for self-constraint by the individual, cooperation among governments, and economic reform toward sustainable use. Of the competing policies, the most formidable is the *Global Biodiversity Strategy: A Policy-makers' Guide*. Because *Genes for Sale* is in direct competition with the *Strategy*, the two must be carefully compared for their expected efficiency and equity in achieving conservation.

The *Strategy* is a user-friendly booklet that succinctly presents the prob-

Biophilia: Where Are You?

Iguaçu is a magical place. The waters of the vast Rio Paraná dump into a narrow canyon, making it one of the most thunderous falls in the world. The views are breathtaking from both the Brazilian and Argentine sides—a beauty truly indescribable. However, the Argentine side is a bit more beautiful. Argentina has an elaborate trail system that leads one right into a place called, quite appropriately, the Throat of the Falls. One passes through panoramic vistas and wetland habitats full of colorful butterflies, toucans, and even monkeys. Indeed, no one should go to Iguaçu without crossing the border, a drive of 18 kilometers, and seeing the Argentine side. At this writing, the cost was only about $12.

On my Fulbright lecture tour of Brazil, I stopped at Iguaçu and asked many residents of the town whether they had ever been to the Argentine side. Many gestured by shaking their index finger and, at the same time, making a clicking sound that emphatically means *no*. Although $12 is a significant amount of money, it is probably not prohibitive to the working-class residents of Iguaçu.

This lack of interest in the environment is not peculiar to the residents of Iguaçu. I witnessed it among the middle classes wherever I went. With respect to the Amazon, almost no one had seen it; yet many had seen Disney World and the shopping malls of Miami. One can only conclude that most Brazilians place a low value on experiencing the aesthetic beauty of the Amazon. The sad truth seems to be that biophilia may be too weak to drive conservation.

lem of mass extinction and offers a complex 85-point guideline for action. Inasmuch as population growth and climate change are divorced from the actions and yet are the greatest threats to biodiversity in the long run, the guideline does not provide a sustainable solution. But the absence of population and climate components is not the fatal flaw of the *Strategy*. What defeats the *Strategy* is that many of its specific actions are so sweeping that a policy maker would have no idea as to the relative impact of each on the goal of conserving biodiversity.

In contrast, the thesis of this book is that intellectual-property rights are the key to conservation. So the brief allusion to intellectual-property rights in action 63 of the *Strategy*, "Improve and expand legal mechanisms to protect species," is grossly underemphasized. To discover the mechanisms, the reader must turn to action 18; "adopt in 1992, the international Convention

on Biological Diversity." The part of the convention that deals with intellectual-property rights is Article 16, paragraph 1:

> Each Contracting Party, recognizing that technology includes biotechnology,
> and that both access to and transfer of technology among Contracting Parties
> are essential elements for the attainment of objectives of this Convention,
> undertakes subject to the provisions of this Article to provide and/or facili-
> tate access for and transfer to other Contracting Parties of technologies that
> are relevant to the conservation and sustainable use of biological diversity
> or *make use of genetic resources* and do not cause significant damage to the
> environment.[11]

Who are the contracting parties? The answer, of course, is governments.
According to the convention, governments of the North will have to acquire
biotechnology from their industries and then transfer it, on favorable terms,
to the governments of the South. The thinking is: the North owns biotechnology, and the South owns biodiversity; through trade, both can be better
off. The trouble here is twofold: it is often individuals within the South and
not their governments who control biodiversity (recall the "paper parks"),
and the transfer of biotechnology may be of no benefit to the individuals
who control the biodiversity even if it were passed on to them (a bartered
good is a poor substitute for money).

The biotechnology lobbies of the North understood immediately how
Article 16 of the convention could compromise their ownership over intellectual property. The compromise was so obvious that the vigorous opposition of the lobbyists should have been no surprise to the drafters of the convention (inter alia, World Resource Institute, the International Union for the
Conservation of Nature, and the United Nations Environment Programme).
What was surprising was how many countries *did* sign the convention; only
the United States weighed the objections more heavily than the political benefits that would accrue with signature. Despite the urging of the head of the
EPA, William Reilly, to sign the convention, President Bush chose not to sign.
With that decision, the United States remained isolated and unpopular.

The refusal of any country to sign, much less the United States, is no small
matter. The biotech industry in the country that refuses gains a comparative
advantage over competing industries in the signatory countries. The reason
for the advantage is the free-rider problem: the industry in the nonsignatory
country can still enjoy access to biodiversity and yet not have to pay the
costs associated with its preservation. In such a situation, one would expect
resentment to build in the signatory countries until their respective governments rescinded the convention.

The position of the United States did not really change with the election of Bill Clinton in 1992. Four months after taking office and on the eve of Earth Day 1993, Clinton fulfilled a campaign promise and signed the convention. But with respect to Article 16, Clinton said he would attach an "interpretative statement."[12] He told reporters, "We think we have done the work necessary to protect the intellectual property of American companies that they would not have to share technology with developing countries that provide resources for products manufactured by those companies."[13]

Without intellectual-property rights over natural genetic information, the convention will fail to achieve an efficient and equitable allocation of land use in the South. With this in mind, *Genes for Sale* is offered as an alternative to the *Strategy* and the convention it endorses. For purposes of comparison, *Genes for Sale* has been reduced to 10 principles that can compete with the Ten Principles for Conserving Biodiversity found in the *Strategy*. Together they should be weighed for their expected efficiency and equity at achieving conservation.

The ten principles for conserving genetic information are

1. Drop the terms "biodiversity" and "sustainable development"[14] from your vocabulary. The former is illogical, and the latter violates the entropy law. Replace the word "biodiversity" with the term "genetic information," and "sustainable development" with this baseline ethic: *those who benefit, pay the costs associated with the benefit, and those who generate a cost, pay for that cost.*

2. Endorse legislation giving equal protection to artificial and natural information. At the same time, attenuate the ability to alienate the new property rights. Leases should be no longer than the term of the intellectual-property protection and transfers no greater than 50 percent of the property.

3. Endorse a temporary general environmental fund (GEF) for conservation funded through royalties. The first task of the GEF should be to integrate existing biological and land-title inventories into one central database. Under no circumstances should the World Bank assume responsibility for the GEF; it has a proven record of failure.[15] NGOs and agencies with proven records of success should assume responsibility.

4. Remit part of the royalties to owners of habitats typical of the organism in which the useful genetic information was discovered and proportional to the landowner's relative share of that habitat. Remit the other part to fund research on the application of

Ten Principles for Conserving Biodiversity*

1. Every form of life is unique, and warrants respect from humanity.
2. Biodiversity conservation is an investment that yields substantial local, national, and global benefits.
3. The costs and benefits of biodiversity conservation should be shared more equitably among nations and among people within nations.
4. As part of the larger effort to achieve sustainable development, conserving biodiversity requires fundamental changes in patterns and practices of economic development worldwide.
5. Increased funding for biodiversity conservation will not, by itself, slow biodiversity loss. Policy and institutional reforms are needed to create the conditions under which increased funding can be effective.
6. Priorities for biodiversity conservation differ when viewed from local, national, and global perspectives; all are legitimate, and should be taken into account. All countries and communities also have a vested interest in conserving their biodiversity; the focus should not be exclusively on a few species-rich ecosystems or countries.
7. Biodiversity conservation can be sustained only if public awareness and concern are substantially heightened, and if policy-makers have access to reliable information upon which to base policy choices.
8. Action to conserve biodiversity must be planned and implemented at a scale determined by ecological and social criteria. The focus of activity must be where people live and work, as well as in protected wildland areas.
9. Cultural diversity is closely linked to biodiversity. Humanity's collective knowledge of biodiversity and its uses and management rests in cultural diversity; conversely, conserving biodiversity often helps strengthen cultural integrity and values.
10. Increased public participation, respect for basic human rights, improved popular access to education and information, and greater institutional accountability are essential elements of biodiversity conservation.

*Global Biodiversity Strategy: Policy-makers' Guide (Washington, D.C.: World Resource Institute, The World Conservation Union, United Nations Environment Programme in consultation with Food and Agriculture Organization and United Nations Education, Scientific and Cultural Organization, 1992), 21.

artificial intelligence to taxonomy. When accurate and cheap taxonomic techniques are fully developed and land-title inventories are complete, remit royalties directly to the landowners on whose property the genetic information is stored.

5. Resist giving or receiving money for environmental protection. If you give, remember that charity is not sustainable. Eventually you will learn to say *no*.[16] And if you receive, remember this Eskimo proverb "Gifts make slaves, as whips make dogs."[17]
6. Endorse family planning on a massive scale. Free Norplants for every woman and condoms for every man.
7. Endorse a greenhouse permit accord with harsh penalties for recalcitrant countries *if and only if* you find the climatological evidence persuasive.
8. Endorse legislation that decouples northern finance from southern dictatorships.
9. Endorse mechanisms that guarantee human rights.
10. And for southern politicians, especially Brazilians, be on military alert for northern aggression on the pretense of human-rights abuse.

The technical expert and political adviser may say, hey, wait a minute, where did items 8, 9, and 10 come from? Are you pulling a fast one? Slipping in your own cause célèbre? Item 8 entered tangentially in Chapters 8 and 9. Decoupling finance from dictatorship is a nascent movement in the United Kingdom and must not be stillborn. If something similar had been endorsed 30 years ago by the democratic governments of the North, it is highly doubtful that the South would have ever suffered the current debt crisis. The relationship between the debt and the deforestation needs no review in this conclusion.

Items 9 and 10 seem sensational. They enter this conclusion by way of amplification effects, boundary conditions, and bifurcation points—the core concepts of NET. A small, insignificant event, given the right boundary conditions, can amplify and cascade. This is often the story of war, and there is no shortage of examples, some of which are even tragi-comic. For example, King Louis Philippe of France went to war with Mexico over a croissant. History records it as the Pastry War. Oh so what, you say, that was the nineteenth century! Well within recent memory, Honduras fought a war with El Salvador over a football match.[18] Today in Brazil there are dangerous boundary conditions that could easily amplify into war given some unforeseeable bifurcation point. To list just three: the authorized violence in rural areas, the complicity of police in the extermination of street children,[19] and the extinction of Amerindian tribes. Any one of the three is a perfect excuse for foreign intervention and the resultant internationalization of the Amazons.

Not Even Wrong

The Third World debt crisis has many fathers. Most are American. In the 1970s, Secretary of State Henry Kissinger prided himself on recycling petrodollars from Saudi Arabia to Latin America via U.S. banks. For a fast buck, money was lent pell-mell to tin-pot dictators. So what if much of the money was subsequently stolen by the dictators? Much of it was redeposited in the very banks that made the loans! The loans were secured by government bonds; whole countries would have to default in order for any individual loan to sour. Of course, that is exactly what happened.

To keep power, the opulent dictators ruled not with an iron fist, but with an invisible hand. Nowhere was this more true than in Argentina, where 15,000 Argentines went missing (*desaparecidos*). The lucky ones were tossed out of planes and into the lakes and rivers; the unlucky ones were first tortured and then buried in unmarked graves. The parents of the missing, some 20 years later and now pensioned, are exhuming their children with the faint hope that forensic analysts can sort out the bones. The parents want a proper burial.

Did the American government know of the missing? Not only did it know of the torture, but it routinely apologized for the dictators! Former American ambassador to the United Nations Jeane J. Kirkpatrick would often go before the television cameras and try to sell the idea that "totalitarian" and "authoritarian" governments were in no way the same. She would claim that Fidel Castro was "totalitarian" and we should have nothing to do with him; but since the generals of Argentina were "only" authoritarian, we could conduct business as usual.*

It is easy to say that Kirkpatrick was simply mistaken in her political theories. To the bereaved and aged parents, she was not even wrong; she represented a government complicitous in the murders of their sons and daughters.

Other than assigning blame where blame is due, what else can be done? The first measure is for each OECD country to adopt legislation similar to the newly enacted Torture Victim Protection Act of the United States.† The act extends U.S. jurisdiction for crimes of torture and summary execution regardless of where the crime is committed. The next measure is to link democracy, the ultimate human right, to finance. Lynda Chalker, Minister of Foreign Development in the United Kingdom, is suggesting that Britain not lend to Third World governments that violate human rights. Minister of Foreign Affairs Douglas Hurd has urged the other countries of the European Union to do likewise.

*See Jeane J. Kirkpatrick, "Dictators and Double Standards," *Commentary*, November 1979, 34–45, and *The Reagan Phenomenon: And Other Speeches on Foreign Policy* (Washington, D.C.: American Enterprise Institute, 1983).

†Public Law 102-256, *United States Statutes at Large*, vol. 106, 73.

The cynic may say: So what? So what if the Amazon is internationalized? Internationalization is a political issue, not a conservation issue. The cynic could not be more wrong. Internationalization means centralization, and centralization does not work. Centralization would fail because the bureaucrats who would be managing the Amazon would not be the ones living in the Amazon. Centralization would violate the efficiency criterion of property right analysis, *those who control an asset should be the ones who derive the benefits from that asset.*

The threat of internationalization is not far-fetched: on the eve of the Hague Conference in 1989, President François Mitterrand of France advanced a frightening thesis: sovereignty is no shield for environmental aggression. The thesis is frightening in light of NET. From NET, we know that all life is entropic and nothing is sustainable. Because everyone must threaten the environment to some degree, the Mitterrand thesis implies a carte blanche for a northern land grab any where and at any time. For these reasons, items 9 and 10 are very much the legitimate domain of *Genes for Sale.* Brazilians must think tactically and ask themselves: Why give Northerners powerful rhetorical ammunition? With just the three aforementioned human-rights abuses (and the list could go on), an American president could whip up a xenophobic frenzy at home, amass the troops, and find easy pickings in the Amazons.[20] This is the lesson to be drawn from the American–Iraq war.[21] With respect to Brazil, the only logical conclusion to draw is that the United States will intervene if (1) it becomes convinced that the Amazons contains something it wants, and (2) if the ethical excuse (e.g., the authorized violence in rural areas, the extermination of street children, the extinction of Amerindian tribes) can pass some litmus test of public opinion at home. Why give Northerners this powerful rhetorical ammunition? Why risk the internationalization of the Amazons? Installing mechanisms to safeguard human rights becomes a legitimate form of defense.

Installing mechanisms to safeguard human rights will bring in its train other complex issues that must be addressed now. Chief among them is indigenous lands. Human rights for native peoples inevitably leads to self-determination. And self-determination will mean greater property rights over habitat in the homelands, the communities, and the reservations. Many environmentalists may think that this will be good for conservation. They could not be more wrong. Self-determination can lead to a flouting of federal environmental laws under the guise of independent jurisdiction—again, the Mitterrand thesis. Nevertheless, self-determination is the only moral course. Fortunately, privatization becomes a way out of the heart-wrenching choice between human rights and environmental protection. As property rights are

extended to genetic information, native peoples will have an incentive to preserve the habitat in their sovereign homelands, communities, and reservations.
Native peoples could even become quite rich from the royalty stream.
The causation that runs from human rights to environmental degradation
is no armchair abstraction. It is quite real. Throughout the North, native
peoples are demanding some degree of self-determination and getting con-
cessions instead. Let me give just one example that is not atypical. A bitter
dispute now rages on Kodiak Island, Alaska.[22] The island is home of the
largest land carnivores on earth, the Kodiak bear. Years ago, the U.S. government created a 0.8-million-hectare National Wildlife Refuge to protect
not only the bear but also 200 pairs of bald eagles, 200,000 waterfowl, and
2 million winter sea birds. But there are also 4,000 native Kodiak Alaskans
on the island. These natives live in a Green poverty; they want an end to
it—they want to develop the island. They believe they have the right to do
so under the Alaska Native Claims Settlement Act of 1971. The act dispersed
land within the refuge in exchange for resolving, once and for all, the ancestral claims. The natives read this settlement as entitling them to full land
use: commercial canneries, logging mills, airstrips, fishing camps, and hunting lodges. The federal authorities see it as only traditional uses: hunting
and fishing. With an adroit public-relations campaign and just the right rhetoric, the natives might even be able to swing some form of self-determination and thereby secure full land use. Therefore, a heart-wrenching choice
has emerged between human rights and environmental protection. The only
way out is to extend property rights to the genetic information of the habitat. In a world that provides equal protection to artificial and natural information, these native Alaskans will collect handsomely for the commercialization of the information embodied in the bears, the eagles, and the
waterfowl of Kodiak Island.

Every good salesman knows that you can oversell the product. Because
Genes for Sale invites so many complexities, I could easily go on and on.
But to do so I would hazard oversale. Recall the first sentence of the first
chapter of this book: economics is a rhetorical enterprise. This book is a
piece of rhetoric and, as such, should not go beyond the optimal point of
persuasion. I believe I am rapidly reaching that point. Perhaps the trip to
Kodiak Island even passed it. So on that note, I abruptly end the chapter,
the page, this sentence, and the book.

Notes

Dedication

1. Vivienne Rae Ellis carefully documents the life and times of the "Last Tasmanian." She answers the question raised in the very title of her book *Trucanini: Queen or Traitor?* new expanded ed. (Canberra: Australian Institute of Aboriginal Studies, 1981). For a compelling argument that the mixed-blood descendants of the last full-blood Tasmanians should also be recognized as aboriginal, see Cassandra Pybus, *Community of Thieves* (Port Melbourne, Australia: Heinemann, 1991).

   The assertion that Trucanini harbored unique genes is a simple deduction of the 1,000-year hypothesis of gene-culture coevolutionary theory. According to the hypothesis, genes bias the adoption of cultural traits, and those traits in turn act as selective agents for the genes that can best assimilate them. For a rigorous treatment, see Charles J. Lumsden and E. O. Wilson, *Genes, Mind, Culture* (Cambridge, Mass.: Harvard University Press, 1981). The same authors popularized the idea in *Promethean Fire* (Cambridge, Mass.: Harvard University Press, 1983).

2. The discerning reader may perceive a contradiction: How can property rights never be fully delineated and yet Trucanini not have any? Logically, this is indeed a contradiction. But practically speaking, one must begin any measurement somewhere. In the continuum of property rights, Trucanini was at the very start even though, theoretically, she had some such rights. See, for example, Robert Hughes, *The Fatal Shore* (New York: Knopf, 1987), 422: "Trucanini, one would presume, had every reason to hate the whites. In fact she sought their company . . . selling herself for a handful of tea and sugar." Prostitution for a handful of tea and sugar is approximately zero property rights.

3. The aboriginal movement to regain property rights is gaining momentum. Even small groups like the 1,000 impoverished Micmacs in Maine are winning recognition of ancestral claims. See Lawrence Lack, "Micmac Indians Seek to Reclaim Their Past," *Christian Science Monitor*, 26 September 1990, 9. Nowhere

is the claim more sweeping than in British Colombia, where some 90 percent of the land is potentially under dispute. See Fred Langan, "British Columbia's Indian Claims," *Christian Science Monitor*, 21 November 1990, 6. And in Australia, not a week passed in 1993 without a front-page article regarding the ruling of the High Court—the Mabo decision—which enables aborigines to sue for title over lands for which they can substantiate a continuous connection.

4. Sampling the genetic information of indigenous races before they become extinct is the goal of the Human Genome Diversity Project, an ambitious effort vigorously supported by world-renowned geneticists. See, for example, Luigi Luca Cavalli-Sforza, Paolo Menozzi, and Alberto Piazza, *The History and Geography of Human Genes* (Princeton, N.J.: Princeton University Press, 1994). Although geneticists may sympathize with the plight of endangered peoples, they often see themselves as powerless to do anything. *Genes for Sale* suggests that they are not powerless. Geneticists can support the creation of intellectual-property rights over the genetic information embodied in the indigenous peoples they sample. See Roger Lewin, "Genes from a Disappearing World," *New Scientist*, 29 May 1993, 25–29.

Preface

1. To appreciate the difficulty in assembling just one component part of the mass-extinction crisis, see Nigel Stork and Kevin Gaston, "Counting Species One by One," *New Scientist*, 11 August 1990, 43–47.

2. Roger Sedjo, "Property Rights and the Protection of Plant Genetic Resources," in *Seeds and Sovereignty: The Use and Control of Plant Genetic Resources*, ed. Jack Kloppenburg, Jr. (Durham, N.C.: Duke University Press, 1988), 293–314. Sedjo does not treat genes as information per se but as a "resource." The word "resource" has been rejected in this book because information is a resource with very peculiar properties; direct usage of "information" implies immediately those peculiar properties. See Kenneth Joseph Arrow, *The Economics of Information* (Cambridge, Mass.: Belknap Press, 1984).

3. The capitalization of genetic information is analyzed in historic detail by Jack Kloppenburg in *First the Seed: The Political Economy of Plant Biotechnology* (Cambridge: Cambridge University Press, 1988).

4. Calestous Juma, *The Gene Hunters: Biotechnology and the Scramble for Seeds* (Princeton, N.J.: Princeton University Press, 1989).

5. Privatization applied to conservation is fairly new but certainly not original. See Ernst Lutz and Herman Daly, "Incentives, Regulations, and Sustainable Land Use in Costa Rica," *Environmental and Resource Economics* 1 (1991): 179–94; Terry L. Anderson and Donald R. Leal, *Free Market Environmentalism* (Boulder, Colo.: Westview Press, 1991); and Jeffrey A. McNeely, *Econom-*

ics and Biological Diversity (Gland, Switzerland: International Union for the Conservation of Nature, 1988). A preemptive strike against any extension of intellectual-property rights to genes is attempted by Pat Roy Mooney's *Seeds of the Earth: A Private or Public Resource?* (Ottowa: Canada Council for International Co-operation, 1980). Mooney practices what he preaches; his book carries no copyright. A good volley between an opponent and an advocate of extension of property rights appeared in *Bioscience*, 37, no. 3 (1987): 215–18: H. G. Wilkes, "Plant Genetic Resources: Why Privatize a Public Good?" and "Jack Kloppenburg, Jr., and Daniel Lee Kleinman Reply." More recent critiques of such privatization can also be found in the last chapter of Fred Pearce, *Green Warriors* (London: Bodley Head, 1991), and in Brian Belcher and Geoffery Hawtin, *A Patent on Life: Ownership of Plant and Animal Research* (Ottawa: International Development Research Centre, 1991). The most rhetorical condemnation of any effort to enclose "the vast interior commons that make up the insides of nature" is found in Jeremy Rifkin, *Biosphere Politics* (New York: Crown, 1991), 65.

For a critique of the intellectual-property-rights mechanisms of the Convention on Biological Diversity, see Vandana Shiva, "Biodiversity, Biotechnology, and Profit: The Need for a People's Plan to Protect Biological Diversity," *Ecologist* 20, no. 2 (1990): 44–47. Also from India is an excellent journal devoted to intellectual-property rights: *HoneyBee* (Indian Institute of Management, Vastrapur, Ahmedabad 380 015, India). The editor, Anil Gupta, places this question on the footer of each page: "Will you stand by the intellectual property rights of peasants?" Gupta suggests a quid pro quo: "India should respect patents and simultaneously apply international patent protection to all of the country's wild and domesticated plant and animal resources" ("Poverty Abounds in Biodiversity-rich Areas," *Down to Earth*, 15 September 1992, 35).

Acknowledgments

1. Most economists and biologists present their ideas in a sterile fashion. The sterility is largely the expression of an inferiority complex relative to physics. To emulate the rigor of physics, economists and biologists have lifted not only its language but also its style (including proofs and topology). Hence, the mimicry of *Physics Review* by the leading theoretical journals in both economics and biology. Nevertheless, over the years, there have been some notable holdouts. In economics, J. Kenneth Galbraith comes to mind; in biology, Steven Jay Gould. Both Galbraith and Gould have presented theoretical issues to the educated public, and both have suffered ridicule for doing so. For the perils of spreading the scientific word, see Bruce Charlton, "The Perils of Popular Science," *New Scientist*, 18 August 1990, 36–40.

Chapter 1

1. Donald McCloskey, "The Rhetoric of Economics," *Journal of Economic Literature* 21, no. 2 (1983): 481–517.

2. Nowhere has privatization been more sweeping than in Brazil, where it has been accompanied by riots and other violence. Upon assuming the presidency in 1990, Fernando Collor privatized huge Brazilian industries, thereby igniting a revolution—not of the metaphorical sort, but a real revolution. For example, with the public auction of the newly privatized steel industry USIMINAS, 600 policemen battled 3,000 demonstrators in Rio. As reported in the newspaper, "the Praça 15 and the streets surrounding the Stock Exchange were turned into an authentic battleground, between 12:30 and 2:00 P.M. Cars were overturned, windows broken, 13 people were detained and more than 70 people, among them 52 policemen and 8 journalists, were wounded by projectile glass and stones as well as tear gas" ("Bozano Leads in the Purchase of USIMINAS and Pays an Extra 15%," *Jornal do Brasil*, 25 October 1991, 1).

   A few days later, the supporters of "privatization" struck back, although not physically. José Assis Simões identified the vested interests among the opponents in "Who Is Against Privatization?" *Jornal do Brasil*, 26 October 1991, 11. Simões lists five groups, the fifth of which has special significance to this book:

   > The last group is that of the hybrids. These comprise among others: the military, the nationalists, the unionists, the conservatives of the right and the conservatives of the left. To justify nationalization, they forge new concepts: "national security," "strategic interests," "reasons of state." Recently they have adopted the slogan: "Nationally owned, the patrimony of the people." In essence the patrimony of the people is that which is in the hands of the people. But the state-run companies are in the hands of the politicians, the technocrats, the bureaucrats, and the unionists. Many of these companies are deep in debt and suffer losses due to negligence, corruption, inefficiencies, and a host of other costs for "the people" to pay. This has become our "patrimony."

3. Sometimes the quest for ownership becomes the principal agent of deforestation. In Brazil, landowners with shaky titles will deforest to demonstrate "improvement" of the land. Rubber tappers whose families have worked the forest for generations are thereby dispossessed. Of course, conflicts ensue and violence appears authorized in the Brazilian interior. In *The Fate of the Forest* (New York: Harper Perennial, 1990), Susanna Hecht and Alexander Cockburn evaluate the perverse laws and inadequate judicial system along with the other agents of deforestation. They conclude that the laws and the system are the principal agents. The solution they propose literally does not appear until the last word of their book: "justice."

4. Adam Smith's most famous words are repeated in virtually every introductory

textbook of economics. The original source is *An Inquiry into the Nature and Causes of the Wealth of Nations* (1776), ed. Edwin Cannan, with an introduction by Max Lerner (New York: Random House, 1937), 11. Ever since Smith, the notion that selfish interests can result in common good has become the mantra of the professionally trained economist. However, the mantra is by no means universally chanted. For example, the sentiment of J. Trevor Williams, former director of the International Board for Plant Genetic Resources (IBPGR), is not unusual among biologists: "These genetic resources have always been regarded by the. IBPGR as a heritage of mankind to be freely available to all bona fide users" ("Identifying and Protecting the Origins of Our Food Plants," in *Biodiversity*, ed. E. O. Wilson [Washington, D.C.: National Academy Press, 1988], 241).

Chapter 2

1. For a report of the Congress, see Laura Tangley, "Biological Diversity Goes Public," *BioScience* 36, no. 11 (1986): 708–15.
2. Michael Soulé, "Mind in the Biosphere," in *Biodiversity*, ed. E. O. Wilson (Washington, D.C.: National Academy Press, 1988), 465–69.
3. A tongue-in-cheek synopsis of the use and abuse of rhetoric in the environmental movement can be found in Paul Spencer Wachtel and Jeffrey A. McNeely, *Eco-Bluff Your Way to Greenism* (Chicago: Bonus Books, 1991).
4. Brian J. Huntley, "Conserving and Monitoring Biotic Diversity," in *Biodiversity*, ed. Wilson, 258.
5. Out of the Earth Summit, RIO'92, emerged a plethora of studies stating the problem and surveying the solutions. Two of the best were the special issues by the Royal Swedish Academy of Sciences: "Population, Natural Resources, and Development," *Ambio* 21, no. 1 (1992); "Economics of Biodiversity Loss," *Ambio* 2, no. 3 (1992).
6. Wilson, "The Current State of Biological Diversity," in *Biodiversity*, ed. Wilson, 16.
7. E. O. Wilson, *Biophilia* (Cambridge, Mass.: Harvard University Press, 1984), 131–32.

Chapter 3

1. Only very recently have economists begun to look to the evolutionary tree for inspiration. The most rigorous analysis has been made by Martin L. Weitzman, "On Diversity," *Quarterly Journal of Economics*, May 1992, 363–405, and "What to Preserve? An Application of Diversity Theory to Crane Conservation," *Quarterly Journal of Economics*, February 1993, 157–183. After several pages of topological proofs in the 1992 article, Weitzman deduces:

When any species becomes extinct, the loss of diversity equals the species' distance from its closest relative, and this myopic formula can be repeated indefinitely over any extinction pattern, because any subevolutionary tree of an evolutionary tree is also an evolutionary tree. When a species becomes extinct, the loss of diversity is calculated as if its evolutionary branch were snapped off the rest of the tree and discarded.

The deduction is consistent with evaluating GCFs *under uncertainty* by looking at the position of genetic information in the evolutionary tree. However, the term "GCF" is more precise than Weitzman's diversity measures, inasmuch as GCF explicitly recognizes that structural distance is not an exact proxy for functional distance. By ignoring this crucial difference, Weitzman renders diversity topologically tractable. But the tractability is bought at the expense of potentially disastrous implications. Extending the examples of the box on page 19, say John Moore had a healthy monozygotic twin and that John Moore's cancerous spleen was owing to a mutation (neither is a heroic assumption: geneticists have well documented that only a few genes can code for such deleterious effects as cystic fibrosis, Huntington's chorea, and sickle cell disease). In light of John's twin, consider Wietzman's "twin property":

> The [identical] twin property is an important statement about continuity of diversity in species addition or subtraction. If a species is added that is very closely related to an existing species, it should only raise the value of diversity by a very small amount that goes to zero in the limit as the added species becomes an identical twin with an existing species. ("On Diversity," 391)

By the identical-twin property, John Moore's tissue could have been eliminated, even though it is worth $3 billion. The rejoinder can easily be anticipated: John twin's is not a true identical twin—after all there are those several mutant genes. However, this expected rejoinder is untenable: to discover the differences, one would need a genetic map, a proposition known as the Human Genome Project with a price tag of several billion dollars. To make matters worse for Wietzman's topological tour de force, there is Motoo Kimura's nagging junk DNA. Two salamander species may be almost identical in morphology, but genetically extremely distant due to the loading of hundreds of thousands of gibberish genes! By the Weitzman criteria for diversity, we might end up saving genetically distant salamanders that are morphologically identical and yet jettison the valuable John Moore.

This lengthy criticism of Weitzman's analysis is hardly abstract or pedantic. It is very real and practical. Under the National Science Foundation Economics Program grant 9122235, awarded on 29 April 1992, Weitzman, as principal investigator, received $145,770 for an abstract that made plain the relevance of his proposed work: "Theoretical applications will include the use of cost–benefit analysis in analyzing conservation issues. . . ." (Abstract Narrative,

National Science Foundation). When one places the cost of the grant in perspective, one sees how poorly U.S. taxpayers were served. The landmark 1987 debt-for-nature swap by Conservation International (CI) and the government of Bolivia was had for only $100,000 and probably saved thousands of species and a huge chunk of the pristine province of Bení. Weitzman's grant cost 50 percent more than the CI–Bolivia swap, and to the extent it legitimizes cost–benefit analysis, will actually *promote* mass extinction.

2. To my knowledge, Jacques Monod was the first to conceptualize DNA as information, in *Chance and Necessity* (New York: Random House, 1972). For the original formulation of the Shannon–Weaver equation, see Claude E. Shannon and Warren Weaver, *The Mathematical Theory of Communication* (Urbana: University of Illinois Press, 1949). For an expansion of the issues Shannon and Weaver pioneered, see Harvey S. Leff and Andrew F. Rex, eds., *Maxwell's Demon: Entropy, Information, Computing* (Bristol: Adam Hilger, 1990). For the relationship between the Boltzmann equation, the Shannon–Weaver equation, and the thermodynamic link or lack thereof, see Jeffrey S. Wicken, *Evolution, Thermodynamics, and Information: Extending the Darwinian Program* (New York: Oxford University Press, 1987). For the use of Shannon–Weaver in ecology, see E. C. Pielou, *Mathematical Ecology* (New York: Wiley, 1977).

3. See the introductory remarks by Weaver, *Mathematical Theory of Communication*, especially p. 8: "two messages, one of which is heavily loaded with meaning and the other of which is pure nonsense, can be exactly equivalent, from the present viewpoint, as regards information." However, one must apply this point with much caution: it would be a grave mistake to conclude that a gene whose function is unknown is, therefore, nonsense. Many of these genes may be essential in translation and splicing mechanisms.

4. On the neutral theory of evolution, see Motoo Kimura, *The Neutral Theory of Evolution* (New York: Cambridge University Press, 1983).

5. E. O Wilson, ed., *Biodiversity* (Washington, D.C.: National Academy Press, 1988), vi.

6. See I. M. Copi, *Introduction to Logic* (London: Macmillan, 1966).

7. G. Carleton Ray, "Ecological Diversity in Coastal Zones and Oceans," in *Biodiversity*, ed. Wilson, 36. The meaning of biodiversity was the theme of a meeting of world-class biologists held at the Natural History Museum, London, 17–19 June 1992. See Henry Gee, "The Objective Case for Conservation," *Nature*, 25 June 1992, 639.

8. Regarding John Moore, see Robert Reinhold, "The Ownership of Cells," *U.S. News & World Report*, 23 July 1990, 58; "Ruling Raises Fear of Research Curbs," *New York Times*, 24 July 1988; Joan O. C. Hamilton, "Who Told You You Could Sell My Spleen?" *Business Week*, 23 April 1990, 38; and "Spleen Rights," *Economist*, 11 August 1990, 34–35. John Moore is not alone; a GCF perhaps even more valuable than that of Moore resides in Helen Boley. Boley is a petite 61 year old who may be the only person ever born with two copies of the "Methuselah gene"—perhaps the key to longevity. See Jerry E. Bishop,

"Mom Married Dad, and the Rest May Be Scientific History," *Wall Street Jour-nal*, 5 April 1991, A1, A9.

9. In economic jargon, "uniqueness" means very low substitutability. There is, of course, a continuum in substitutability. Orthodox economists would like to assume that given new technologies and enough time, there is perfect substitutability. Orthodox environmentalists would like to assume the opposite: there are no substitutes—everything warrants conservation. The truth, not surprisingly, is somewhere in between. Take two examples: polyesters and chemotherapeutics. Economists will argue that polyesters are a substitute for natural fibers. But are they perfect substitutes? Definitely not: cotton, silk, and linen are still very much in demand. Environmentalists have argued that the alkaloids from the rosy periwinkle are the only known chemotherapies for childhood leukemia. Maybe so, but substitutes should not be confined to drugs: measures to reduce mutagens and toxins in our environment and prevent childhood leukemia can be substituted for chemotherapies. See Paul R. Ehrlich, "The Limits to Substitution: Meta-Resource Depletion and a New Economic-Ecological Paradigm," *Ecological Economics* 1 (1989): 9–16.

10. To appreciate the ants, see Bert Hölldobler and E. O. Wilson, *The Ants* (Berlin: Springer, 1990). Students of economics will recognize the priceless value of ants for forestry and the total disregard for their legal protection as an example of the diamond–water paradox. Diamonds are inessential; water is essential. Why then are diamonds expensive and water cheap? The answer: the psychological phenomenon of diminishing satisfaction and the relative supply of each. To cut back 1 liter of water from the 1,000 some liters consumed each day by a metropolitan Northerner is not painful; the thousandth liter brings little satisfaction. Water is plentiful. Diamonds are at the other end of the continuum of satisfaction. Because they are so scarce, most people possess few, and the few possessed bring a great deal of satisfaction. So at the margin—that is, from 999 to 1000—people are not willing to pay much for the last liter of water, while they are willing to pay a great deal for their first diamond—that is, from 0 to 1. Ants are more like water than diamonds.

11. The poaching of bears was a major theme of the Eighth International Conference on Bear Research. The conference was covered in "Boom in Poaching Threatens Bears Worldwide," *New York Times*, 1 May 1990, C1, C11. Perfect substitutability with respect to a GCF is buttressed by the isomorphism of the gall bladders in all the bear species. See Judy Mills, "All Gall Bladders Look Alike," *Audubon Magazine*, July–August 1991, 96.

12. Regarding GCFs at the level of the species (*Zea diploperennis*), see H. H. Iltis, J. F. Doebley, R. Guzmán, M. Pazy, and B. Pazy, "*Zea diploperennis* (Gramineae): A New Teosinte from Mexico," *Science* 203 (1979): 186–88. The value of perennity was calculated by W. M. Hanneman and A. C. Fisher, "Option Value and the Extinction of Species" (Manuscript, California Agricultural Experiment Station, Berkeley, 1985).

    Regarding GCFs at the level of the subspecies (*Solanum melongena*), see

E. Bettencourt and J. Konopka, *Directory of Germplasm Collections 4. Vegetables* (Rome: International Board for Plant Genetic Resources, 1990), 217. For the dollar value of damage to commercial fields of eggplants, see Dennis Allan Schaff, "Hybridization and Fertility of Hybrid Derivatives of *Solanum melogena* and *Solanum macrocarpum* L." (M.S. thesis, Rutgers University, 1980).

13. The taxon is only a guide to uniqueness. One should not abide by it too slavishly. Just as the Tasmanian tiger (*Thylacinus cynocephalus*) is much closer to the freckled mouse (*Parantechnius apicalis*) than either is to its placental namesake, functions of commercial value to industry, like the outward appearances of tigers and mice, can be coded by very different genes. Sweetness is probably the best example. The function can be had from nutritive sugars such as glucose, fructose, and sucrose or from a totally unrelated molecule like the amino acid phenylalanine (NutraSweet). See Michael Ruse, "Biological Species: Natural Kinds, Individuals, or What?" *British Journal for the Philosophy of Science* 38 (1987): 225–42, and Peer Bork et al., "What's in a Genome?" *Nature*, 23 July 1992, 287.

14. Public opposition to eradication schemes is not restricted to the fox. The rusa deer (*Cervus timorensis russa*), introduced into New South Wales, Australia, some 100 years ago, even has an NGO committed to its defense! It made the headline of the *Sydney Morning Herald*, 13 April 1993: "Fury as Rangers Slaughter Park's Lovable Aliens." The role of rhetoric is paramount in the defense of the rusa deer. The NGO spokesman told reporters, "The facts are the deer have been here since 1885. They are a great tourist attraction and cause no trouble. They earned the right to their Australian citizenship."

15. David Ehrenfeld, "Why Put a Value on Biodiversity?" in *Biodiversity*, ed. Wilson, 214. Ehrenfeld's concern dovetails with that of one of the architects of modern cost–benefit analysis, E. J. Mishan: "In our growth-fevered atmosphere there is always a strong temptation for the economist, as for other specialists, to come up with firm quantitative results. In order to be able to do so, however, he finds that he must ignore the less easily measured spillovers. . . . As a matter of professional pride, and of obligation to the community he elects to serve, the economist should resist this temptation" (*Economics for Social Decisions* [New York: Praeger, 1974], 109). E. F. Schumacher goes one step further in *Small Is Beautiful: A Study of Economics as if People Mattered* (London: Sphere Books, 1974). Schumacher identifies cost–benefit analysis as

> a procedure by which the higher is reduced to the lower and the priceless is given a price. It can therefore never serve to clarify the situation and lead to an enlightened decision. All it can do is lead to self-deception or the deception of others; for to undertake to measure the immeasurable is absurd and constitutes but an elaborate method of moving from preconceived notions to foregone conclusions; all one has to do to obtain the desired results is to impute suitable values to the immeasurable costs and benefits. The logical absurdity, however, is not the greatest fault of the

undertaking: what is worse, and destructive of civilization, is the pretence
that everything has a price or, in other words, that money is the highest of
all values. (37–38)

For a general look at the application of cost–benefit analysis to environmental
resources, see Partha Dasgupta, *The Control of Resources* (Oxford: Basil Black-
well, 1982). For an attempt at applying cost–benefit analysis to biodiversity,
see Gardner Brown, Jr., and Jon H. Goldstein, "A Model for Valuing Endangered Species," *Journal of Environmental Economics and Management* 11
(1984): 303–9. For a compelling argument that the burden of proof of benefits
should be shifted from environmentalists to industrialists, see Clem Tisdell,
"Economics and the Debate about Preservation of Species, Crop Varieties, and
Genetic Diversity," *Ecological Economics* 2 (1990): 77–90.

Chapter 4

1. See Yoram Barzel, *Economic Analysis of Property Rights* (New York: Cambridge University Press, 1989).
2. The term "paper park" was coined by P. Fearnside and G. Ferreira, "Roads in
   Rondônia: Highway Construction and the Farce of Unprotected Reserves in
   Brazil's Amazonian Forest," *Environmental Conservation* 11 (1984): 358–60.
3. Economists often dream up theories contingent on so many ridiculous assumptions that they themselves become the butt of ridicule. Jokes abound. A classic
   is the three shipwrecked academicians: a physicist, an engineer, and an economist. They are hungry, and a can of beans washes ashore. How to open it? The
   physicist suggests gathering kindling and lighting a fire; by $PV = nRT$, the can
   will explode. The engineer objects: "How messy, why not find a sharp shell
   and pry off the top?" The economist shakes his head: "You guys have it all
   wrong, assume there is a can opener."
   The joke has become an oral tradition among the many critics of neoclassical economics. Therefore, it is hard to fix a time and place to its birth. I would
   venture to guess the origins lie in the preposterous claims made by Milton
   Friedman in the early 1950s that "a theory cannot be tested by comparing its
   'assumptions' directly with 'reality'" (*Essays in Positive Economics* [Chicago:
   University of Chicago Press, 1953], 41).
4. The analogy of intellectual-property rights over software to intellectual-property rights over genetic information has probably occurred to many people. The
   first example in the literature seems to be Cyril deKlemm, "Conservation of
   Species: The Need for a New Approach," *Environmental Policy and Law* 9,
   no. 4 (1982): 117–28. A more recent exposition of the analogy is Timothy M.
   Swanson, "Economics of a Biodiversity Convention," *Ambio* 21, no. 3 (1992):
   250–57.

All analogies are based on the Coase theorem of economics. The theorem was named by George Stigler, "Two Notes on the Coase Theorem," *Yale Law Journal* 99 (1989): 631–33, in honor of Ronald Coase for his seminal paper, "The Problem of Social Cost," *Journal of Law and Economics*, October 1960, 1–44. Today there are legions of economists who specialize in interpreting the many implications of the Coase Theorem. In *Economic Analysis of Property Rights*, Barzel offers one of the more succinct interpretations: "The Coase Theorem simply states that when property rights are well defined and transacting is costless, resources will be used where they are most valued regardless of which of the transactors assumes liability for her or his effects on the other" (55). The criterion for efficiency in ownership is the following corollary, also by Barzel, "The rights to receive the income flow generated by an asset are a part of the property rights over that asset. . . . The maximization of the net value of an asset, then, involves that ownership or ownership pattern that can most effectively constrain uncompensated exploitation" (5). For a historical perspective on property-rights analysis, see Steven N. S. Cheung, *The Myth of Social Cost*, Hobart Paper 82 (London: Institute of Economic Affairs, 1978). For a historical perspective on property rights themselves, see Barry C. Field, "The Evolution of Property Rights," *Kyklos* 42 (1989): fasc. 3, 319–45.

5. For a quick overview of intellectual-property law and computer software, see David Bender, "Protecting Computer Software," *Trial*, June 1990, 58–61. Penalties against pirates in the workplace are effective deterrents when the software applications are strictly corporate, but not so effective when the applications are strictly personal. The property-rights approach can explain the dichotomy. The thief for the corporation is, at best, only a fractional residual claimant on the incremental profits of his cost-saving piracy. However, if this same individual pirates software for his own personal use, he is the full residual claimant.

Chapter 5

1. Pictographs work. For example, Mexico City is home to thousands of commuters who cannot read or write. The transit authority designates its stops with both letters and symbols: the Zócalo stop is symbolized by a silhouette of the cathedral located on the plaza. Symbolizing a metro stop is easy; symbolizing a new policy is not so easy. It will take a great deal of artistic talent to capture both the opportunities and the restrictions of the policy.

2. There are many horror stories to tell about the World Bank. Most center on its financing of projects that resulted in ecocide and the concomitant genocide. One of the most horrific occurred on the Uatumã River in Brazil, where the bank bankrolled the Balbina hydroelectric dam. As reported by Rogério Gribel, "The Balbina Disaster: The Need to Ask Why?" *Ecologist*, 20, no. 4 (1990): 133–35:

The Balbina dam cost nearly a billion dollars of public money, destroyed 236,000 hectares of primary forest, formed a gigantic lake of shallow stagnant water, killed millions of wild animals, flooded indigenous lands, is causing hunger and illness to riverine people—all for just 80 megawatts of electricity. Even those responsible for the dam now admit that it is a disaster, yet similar projects are still being built and no investigation has been carried out to discover why, in the face of clear scientific warnings, Balbina was ever constructed?

3. Even the staunchly conservative *Economist* minces no words in its reporting of the bank's operation: "Lack of direction is only one of the bank's problems. Another lies in their bloated staffs, murky procedures and lax operation controls" ("Bankers at Bay," *Economist*, 26 June 1993, 16–27). As of this writing, the bank is still considering funding two potential environmental fiascos: the Three Gorges Dam in China and the BíoBío Dam in southern Chile.

4. In the spring of 1992, the Supreme Court considered government "taking" through regulations in the case *Lucas* v. *South Carolina Coastal Council*. Real-estate developer David Lucas argued that new state regulations prohibiting beach-side construction denied him compensation for his investment in the beachfront property. The Court ruled 6 to 3 in his favor. See Linda Greenhouse, "Justices Ease Way to Challenge Land-Use Rules That Prevent Development," *New York Times*, 30 June 1992, A18.

5. Eco-dogma has it that rain-forest-turned-pasture is a lunar karst after five years of grazing. If only it were only true, then the case for conservation would be much easier! Unfortunately, it is only partly true. Much of the high land and savannas in the Brazilian interior can be cultivated for permanent crops such as cacao and coffee. See Ellen B. Geld, "Will Farming Destroy Brazil's Amazon Basin?" *Wall Street Journal*, 13 January 1989, A13. Gild is an author and farmer in Tiete, Brazil.

   "Paper parks" are also a matter of degree. For example, the Pacific yew tree (*Taxus brevifloria*) is being poached from the protected forests of the Northwest. Its bark yields a rare chemotherapeutic alkaloid, paclitaxel, more commonly known as taxol. See Gina Kolata, "Tree Yields a Cancer Treatment, but Ecological Cost May Be High," *New York Times*, 13 May 1991, A1, A13. Much official testimony suggests that the poaching of the yew tree is out of control because the agency entrusted with the protection of the National Forests, the U.S. Forest Service, is rife with corruption. See John McCormick, "Can't See the Forest Through the Sleaze," *New York Times*, 29 January 1992, A19.

6. No government acts by the simple maxim "greatest good for the greatest number." From a property-rights perspective, government itself does not exist! Only individuals exist. Governments are composed of bureaucrats and legislators, each pursuing his own selfish interests given the constraints of his contract. The "greatest good for the greatest number" is a shorthand way of describing

situations where the incentives of the constituency are aligned with those of the bureaucrats and legislators.

7. The calculation of profitability is based on estimates from Norman Myers, *A Wealth of Wild Species: Storehouse for Human Welfare* (Boulder, Colo.: Westview Press, 1983), and from Norman R. Farnsworth and Djaja Doel Soejarto, "Potential Consequences of Plant Extinction in the United States on the Current and Future Availability of Prescription Drugs," *Economic Botany* 39, no. 3 (1985): 231–40. For a technical look at the medicinal properties of plants and their potential, see *Plants: The Potentials for Extracting Protein, Medicines, and Other Useful Chemicals: Workshop Proceedings from the Office of Technology Assessment* (Washington, D.C.: Office of Technology Assessment, 1983).

8. The estimate that "one man decade would be required to enumerate one hectare" was made by T. C. Whitmore, "Total Species Count on a Small Area of Lowland Tropical Rain Forest in Costa Rica," *Bulletin of the British Ecological Society* 17 (1986): 147–49. The statistic is quoted in Harold Mooney, "Lessons from Mediterranean-Climate Regions," in *Biodiversity*, ed. E. O. Wilson (Washington, D.C.: National Academy Press, 1988), 158.

Chapter 6

1. For an overview of the gold rush in the Amazons, see David Cleary, *Anatomy of the Amazon Gold Rush* (Iowa City: University of Iowa Press, 1990).

2. Sociobiologists deduce that the agricultural way of life is toil because humans did not evolve for stoop labor. In the more than 3 million years of human evolution from the earliest hominids, a mere 10,000 have been spent in activities other than hunting and gathering. The agricultural way of life, although far more successful than hunting and gathering, has not been around long enough to select genes that code for pleasure from that way of life. See Lionel Tiger and Robin Fox, *The Imperial Animal* (New York: Holt, Rinehart and Winston, 1971).

3. The term "biophilia" was first used by E. O. Wilson, "Biophilia," *New York Times Book Review*, 14 January 1979, 43, and later in his book *Biophilia* (Cambridge, Mass.: Harvard University Press, 1984). William Hamilton argues the same thing but with no reference to Wilson's biophilia in "Discriminating Nepotism: Expectable, Common, Overlooked," in *Kin Recognition in Animals*, ed. David J. C. Fletcher and Charles D. Michener (New York: Wiley, 1987), 430.

4. For a property-rights analysis of homesteading, see Douglas W. Allen, "Homesteading and Property Rights: Or 'How the West Was Really Won,'" *Journal of Law and Economics* 34 (1991): 1–23.

5. Several books on the rain forest and its destruction have been published in the last few years, many the culmination of a decade or more of painstaking research. The issues are so complex that each author, although dealing broadly with the same subject, brings to it a different perspective. See Alex Houmatoff, *The*

World Is Burning: Murder in the Rainforest (New York: Avon Books, 1990); Adrian Cowell, *The Decade of Destruction* (New York: Holt, 1990); Andrew Revkin, *The Burning Season: The Murder of Chico Mendes and the Fight for the Amazon Rainforest* (Boston: Houghton Mifflin, 1990); and Augusta Dwyer, *Into the Amazon* (San Francisco: Sierra Club Books, 1990).

6. Many modern industrialized nations practiced infanticide before they achieved prosperity. Infanticide is a consequence of poverty. In the Third World, we must assume infanticide is an option often taken and left unrecorded. For an analysis of infanticide across cultures, see Nancy Scheper-Hughes, ed., *Child Survival: Anthropological Perspectives on the Treatment and Maltreatment of Children* (Dordrecht: Reidel, 1987). For an analysis of infanticide within the Brazilian culture, see Nancy Scheper-Hughes, *Death Without Weeping: The Violence of Everyday Life in Brazil* (Berkeley: University of California Press, 1992). For a report on the complex issues surrounding abortion and ultrasound technology in China, see Nicholas D. Kristof, "Peasants of China Discover New Ways to Weed Out Girls," *New York Times*, 21 July 1993, A1, A6.

7. Incentives are not enough to effect optimal family size. Women must also be given the know-how; contraceptives like the five-year subcutaneous contraceptive implant (Norplant) would do wonders for the implementation of the policy. The high cost of Norplant in the United States (approximately $580) is owing to monopoly elements in the provision and delivery of U.S. health care. It should not dissuade policy makers in the developing countries, where a competitive price will be a small fraction of the U.S. price.

8. For a look at having nontaxonomists performing taxonomy, see Roger Lewin, "Costa Rican Biodiversity," *Science*, 23 December 1988, 1637. In the article, a leading authority on restoration ecology in the Third World, Daniel Janzen, endorses the idea of having nontaxonomists perform taxonomy:

> The traditional approach for creating an inventory of a biota—figuring out what you have saved—would involve calling on the time and expertise of international systematists in collecting, identifying, and classifying specimens, of about a million species in this case. This would just bring more misery for everyone, because most systematists' resources are already grossly overextended. Instead in Costa Rica, the collection and initial sorting is being done by farmers, other rural workers, and university students, many of whom previously worked in the park areas in more traditional jobs.

9. The Australian government recognizes that the removal of nonindigenous species is a labor-intensive activity. Crean, the minister of primary industries, is advocating a federal jobs-creation program. See *Pest Animals in Australia* (Canberra: Australian Government Publishing Service, 1992). It is also widely accepted that endemic species left in the wild must be manipulated. The rationale is simple: populations of many species have shrunk to such small numbers that, left on their own, a disease or even just a bad storm could reduce

their numbers to that of a nonviable population. Through inbreeding, those that survive would become "the living dead." See D. H. Janzen, "The Future of Tropical Ecology," *Annual Review of Ecology and Systematics* 17 (1986): 304–24, and Michael E. Soulé and Bruce A. Wilcox, *Conservation Biology* (Sunderland, Mass.: Sinauer, 1980).

Chapter 7

1. Land is often improved before title is clear. For example, when an individual puts up a fence without surveying the property, he risks a neighbor taking him to court; the judgment is almost always to remove the fence. When a nation does this, it risks a border war and loss of land improvements. With respect to the latter, the example that comes to mind is the high-rise luxury hotel in Taba, Eilat, Israel. With the return of the Sinai to Egypt, the hotel in Taba is now across the border and belongs to Egypt. See Joel Brinkley, "A Sandy Corner of Egypt Sadly Misses Its Israelis," *New York Times*, 28 February 1990, A4.

2. Boundaries are easy to establish for land traversed by rivers. But what about mountains? The Pyrenees between Spain and France seem like a good boundary. But the jagged mountains themselves have value and can be used as pasture. When one is at the foot of the mountains it is clear whether one is on the Spanish or the French side, but what about when one is on the mountain? How is the line to be drawn across the mountains? The answer that emerged in medieval times is elegant in its simplicity. If rain flows down the mountain and into Spain, that is the Spanish side of the mountain; if it flows into France, that is the French side. Even this simple mechanism only lessens the transaction costs of negotiation; over time there will be soil erosion and the course of the water will reroute. Under black letter law, disputes will arise. Again, we see the basic principle of property-rights analysis: *property rights are never fully delineated*.

3. The term "commons" is used almost as loosely by economists as it is by the public. In economics, "commons" should not mean unowned resources, but assets controlled by a specified group. S. V. Ciriacy-Wantrup and R. C. Bishop make this point in "'Common Property' as a Concept in Natural Resource Policy," *Natural Resource Journal* 15, no. 4 (1975): 713–27.

4. The commons of genetic information may not always be among neighbors. For example, in the United States, the scrub jay (*Aphelocoma coerulescens*) is found on two widely separated properties. There is a population along the Florida gulf coast and another across the continent in California.

5. In the late nineteenth century, Brazil experienced relatively low transaction costs in its control of the genetic information of several species in the genus *Hevea*. Because Brazilians controlled the Amazon River and the river was the gateway to *Hevea*, Brazil was able to exclude nonpaying users. By the turn of the century, Brazil had achieved a virtual monopoly over the GCF of *Hevea*—that is, rubber. The splendor of the opera house in Manaus bears witness to the value

of sole ownership of a GCF. But the monopoly was not to last. Information is just too easy to steal! The British did just that. Sir Joseph Hooker, director of Kew Gardens, persuaded the colonial government of India to sponsor the explorer and naturalist Henry Wickham to get *Hevea* seeds and saplings out of the Amazon. The success of Wickham and subsequent success of rubber plantations in India caused the collapse of the Amazon economy. The squalor of twentieth-century Manaus bears witness to the loss of monopoly privileges. In 1920, Wickham was knighted in honor of his theft some 50 years before. See Warren Dean, *Brazil and the Struggle for Rubber* (New York: Cambridge University Press, 1987).

6. Regarding protocols in communication networks, the argument is basically one of standards. If government sets standards, then cooperation is facilitated and economies of scale emerge. However, once the standard is set, it is hard to go back. The example economist Stan Liebowitz likes to use is the QWERTY keyboard. It has often been claimed that a different configuration of the alphabet on the keyboard would cause less finger fatigue. But to adopt a new standard is costly in terms of both physical and human capital. So, the argument goes, we often become locked into a QWERTY as an accident of history. See S. J. Liebowitz and Stephen E. Margolis, "The Fable of the Keys," *Journal of Law and Economics* 33 (1990): 1–25. Similar arguments can be made for emerging standards in information technologies. The lesson to be drawn is that we must, at the outset, choose our standards carefully. And if we do not? Phyllis Sokol offers the advice of a Turkish proverb in the dedication of her userfriendly book *EDI: The Competitive Edge* (New York: McGraw-Hill, 1989): "No matter how long you have gone on a wrong road, turn back."

7. The analogy of the tolerable inefficiencies of a central bank and tolerable inefficiencies of a proposed gargantuan database occurred to me after reading Milton Friedman and Rose Friedman's account of the stupidities of the U.S. Federal Reserve Bank during the Great Depression of the 1930s: *Free to Choose* (Harmondsworth: Penguin, 1979), chap. 3. Despite the inefficiencies that accompany centralization, the Friedmans do not advocate the abolition of central banking.

8. S. A. Marshall, "Could There Be a New Insect in Your Backyard?" *American Biology Teacher* 52, no. 2 (1990): 101.

9. A species count is a record of what species are on a property. These counts are reported per 0.1 hectare (0.001 km^2). With the majority of tropical species unclassified, total counts have taken place only in the most studied reserves. See Harold A. Mooney, "Lessons from Mediterranean Climate Regions," in *Biodiversity*, ed. E. O. Wilson (Washington, D.C.: National Academy Press, 1988), 157–65. For example, a temperate-zone forest in Missouri will exhibit 21 species, while an area one-tenth the size in Costa Rica will exhibit 236 species. From Mooney's data, one infers that the mean number of species increases as one goes from dry tropical to wet tropical and from temperate to tropical zones. One must also keep in mind that the 90 percent of unclassified species are not evenly distributed: Australia and Brazil are roughly the same geographic size,

but Australia has only some 4,000 unclassified species, while Brazil may have more than 4 million.

10. For a look at biological surveys, see Leslie Roberts, "Hard Choices Ahead on Biodiversity," *Science*, 30 September 1988, 1759–61.

11. "Digitize" may not be the right word or the right technology. Brazilian mathematicians are inventing technologies that interpret images as fractal equations. The equations render a resolution far greater than digitization.

12. Regarding the application of artificial intelligence to taxonomy and related information technologies, see Kenneth Estep, Arne Hasle, Lena Omli, and Ferren MacIntyre, "Linnaeus: Interactive Taxonomy Using the Macintosh Computer and HyperCard," *BioScience* 39, no. 9 (1989): 635–38.

13. Secondary-school teachers of biology are the most obvious supply of parataxonomists. However, others exist. For example, at the June 1993 meeting of the Society for Growing Australian Plants in Cairns, Far North Queensland, professional botanists gathered around a plumber who systematically identified specimens collected by another member on a mountaintop near Cape Tribulation.

14. Richard H. Zander, "TAXACOM, an Online Service for Systematic Botany," *BioScience* 37, no. 8 (1987): 616–18; Karen M. G. Howell, "Federal Government Applications of Write-Once Read-Many (WORM) Optical Disk Systems," *Library Hi Tech News*, January 1988, 1, 2, 7–9; Judith Paris Roth, ed., *Converting Information for WORM Optical Storage* (Westport, Conn.: Meckler, 1990).

15. Molecular biology will eventually replace traditional taxonomic analysis for two simple reasons: (1) it is already better, and (2) it will soon become cheaper as capital substitutes for labor. Molecular approaches are possible because genetic information scrambles at a fairly constant rate. This means that the lower the variance between the genes of two organisms, the closer the phylogenetic relationship. In contrast, traditional taxonomy based on morphologies can easily mistake convergent morphologies as homologous—that is, originating from the same common ancestor. Radioimmunoassay (RIA) can therefore resolve many old taxonomic debates regarding homology. The beauty of RIA is that it can even be done with the small samples of tissue from fossil specimens. See Jerold M. Lowenstein, "Molecular Approaches to the Identification of Species," *American Scientist* 73 (1985): 541–47. Lowenstein notes that "although it should not be counted the sole infallible means of investigation, RIA is the first technique to extract useful genetic information at the subvisible, molecular level, from fossil proteins. In addition to providing estimates of species relationship and evolutionary divergence, it is increasingly finding application in animal and plant taxonomy and in the identification of fragmentary archaeological specimens" (547). New techniques of taxonomy at the molecular level include polymerase chain reaction (PCR). See Norman Arnheim, Tom White, and William E. Rainey, "Application of PCR: Organismal and Population Biology," *BioScience* 40, no. 3 (1990): 174–82.

Returning to John Moore in the box on page 19, one finds a good example of the need for molecular analysis at the level of the individual. Suppose that John won his lawsuit over that $1 billion spleen; suppose also that John's double surfaced in Argentina and claimed to be the long-lost monozygotic twin! The tabloids are full of doubles who look just like Richard Nixon, Jimmy Carter, and Ronald Reagan. How to determine whether that double is indeed the twin of John Moore and not an imposter? Obviously, molecular analysis.

Chapter 8

1. Legislation to the same effect was also passed in New Zealand in the same year. See C. Brown, "Protecting Plant Varieties: Developments in New Zealand," *Victoria University of Wellington Law Review* 83 (1988): 18.

   The Prime Minister's Science Council of Australia recognized the natural wealth of the country in its opening sentence to *Scientific Aspects of Major Environmental Issues: Biodiversity* (Canberra: Australian Government Publishing Service, 1992): "Australia has a rich and unique flora and fauna. We have about 450,000 species or 7% of the world's 6 million species of plants and animals. . . . This is more than twice the combined number of species in Europe and North America north of Mexico. We are one of the seven mega-diversity regions of the world." Australia and New Zealand are also the only biodiverse-rich countries that respect international intellectual-property rights and routinely pay several times more in royalties on foreign-owned intellectual property than they receive from Australian and New Zealand intellectual property overseas (as the current accounts of the respective balance of payments woefully demonstrate). By having respected intellectual-property rights for so many years and at such great cost, both Australia and New Zealand can now take the high moral ground and advocate privatization as an efficient and equitable policy to conserve biodiversity.

   The wheels are already in motion. The state government of Queensland announced its intent to introduce legislation to patent natural genetic codes. See Madonna King, "State to Patent Nature's Secrets," *Australian*, 14 April 1993, 1. Western Australia is moving in the same direction and asserting rights over the "HIV" plants of the genus *Conospermum*. See Susan Katz Miller and Leigh Dayton, "Australia Takes Tough Line on 'HIV Plant,'" *New Scientist*, 3 July 1993, 4.

2. Technically, the higher the price, the more people are priced out of the market for a particular drug and the more elastic becomes the demand. What this means is that pharmaceutical houses will raise their prices until they are in the elastic region at the price that yields maximum profit. The royalty is analogous to a tax. For a graphical exposition of tax incidence, see Robert S. Pindyck and Daniel L. Rubinfield, *Microeconomics* (New York: Macmillan, 1989), 322, fig. 9.20.

3. Regarding *Diplogottis campbellii*, see *The Rainforest Legacy: Australian National Rainforest Study* (Canberra: Australian Heritage Commission, 1987), 100. The book supports the radical notion that ordinary people can make extraordinary scientific discoveries.

4. The pharmaceutical industry has a tainted history in Latin America. For years, the *Physician's Desk Reference* (*PDR*) added indications to the Latin American edition of the publication and omitted the contraindications found in the American edition. The economic effect of the Latin American *PDR* was more pills sold. The health effect of this double whammy was literally overkill—in the case of antibiotics, thousands of deaths due to aplastic anemia. To top it off, the industry segmented each market and exacted a monopoly price. Pills in impoverished Latin America were often 20 times more expensive than the same pills in the United Kingdom. See Milton Silverman, *The Drugging of the Americas* (Berkeley: University of California Press, 1976).

5. The controversy over AZT has gone beyond price gouging to questions of patent ownership and government funding of the basic research. See Joseph Palca, "Monopoly Patents on AZT Challenged," *Science*, 7 June 1991, 1369. Although the challenge has failed in the courts, the publicity surrounding the case indicates public hostility toward the pharmaceutical industry.

6. To counter any talk of price controls on prescription drugs, the pharmaceutical industry lobby is groveling for populist support. It has even sought out the Reverend Jesse Jackson's National Rainbow Coalition. See Robert Pear, "Drug Industry Gathers a Mix of Voices to Bolster Its Case," *New York Times*, 7 July 1993, A1, A12.

7. David Ehrenfeld sees the incentive structure of pharmaceutical research laboratories as the primary reason for the general lack of interest in biodiversity: "pharmaceutical researchers now believe, rightly or wrongly, that they can get new drugs faster and cheaper by computer modelling of the molecular structures they find promising on theoretical grounds, followed by organic synthesis in the laboratory using a host of new technologies, including genetic engineering. There is no need, they claim, to waste time and money slogging around in the jungle. In a few short years, this so-called value of the tropical rain forest has fallen to the level of used computer print-out" ("Why Put a Value on Biodiversity?" in *Biodiversity*, ed. E. O. Wilson [Washington, D.C.: National Academy Press, 1988], 213). The computer manipulation of drugs in research laboratories is also reported in Sue Shellenbarger, "Lilly's New Supercomputer Spurs a Race for Hardware to Quicken Drug Research," *Wall Street Journal*, 14 August 1990, B1, B3.

8. The New York Botanical Garden, in cooperation with institutions in developing countries, has signed agreements with the National Cancer Institute for screening medicinal plants. For any resultant patented drug, royalties will remit to the institutions in the developing country of origin. Lisa Conte, founder of Shaman Pharmaceuticals, has also entered into similar agreements. See Deanna Hodgin, "Seeking Cures in the Jungle," *Insight*, 7 October 1991, 30–31.

9. Not all would agree that the ERA is simple and just. President Bush asked television evangelist Pat Robertson to share God's thoughts on the subject during
the 1992 Republican National Convention. "The feminist agenda is not about
equal rights for women," Robertson said. "It is about a socialist, antifamily political movement that encourages women to leave their husbands, kill their children, practice witchcraft, destroy capitalism, and become lesbians" (James
Adams,"Baker's Storm Focuses Party on Re-election," *Australian*, 31 August
1992, 8).
10. Regarding original intent, see Harry N. Scheiber, "Original Intent, History, and
Doctrine: The Constitution and Economic Liberty," *American Economic Review*,
May 1988, 140–44. Regarding deconstruction, see Christopher Butler, *Interpretation, Deconstruction, and Ideology* (Oxford: Clarendon Press, 1984), and
Jonathan Culler, *On Deconstruction* (Ithaca, N.Y.: Cornell University Press, 1982).
11. Ironically, the Court's reasoning in the landmark *Diamond* v. *Chakrabarty*, 206
U.S.P.Q. 198 (1980) is decidedly Marxist, yes Marxist! Under Marxian labor
theory of value, raw materials are common property. Compensation is only for
work expended in adding value to them. The National Science Foundation is
now stretching "compensation for work expended" to the max. In 1991, it requested patents on 340 human genes whose function is still unknown, the argument being that compensation would be for the work expended in the discovery of the sequence.
12. Quoted in John Maier, Jr., Interview with Gilberto Mestrinho, *Time*, 16 September 1991, 22–23.

There is an ironic twist in the extension of intellectual-property rights to natu-
ral information; the ideology corresponds to that of conservatives, and yet the
tangible benefits are clearly populist. Nowhere is this more evident than in
Brazil. For example, on 8 August 1991, the lead articles in the newspaper *Jornal
do Brasil* touched on themes of intellectual-property rights and privatization:
"Americans Complain About Brazil," 1, 9, and "Darcy Leaves the Senate and
Attacks Privatization," 1, 7. The first headline reported the meeting of Vice
President Dan Quayle with Brazilian President Fernando Collor: "Quayle and
Collor talked about pharmaceutical and information related patents, themes
under discussion in the Congress" (9). The second headline reported the resig-
nation from the senate of the senior statesman Darcy Ribeiro, a man of the left.
In his resignation, Ribeiro vigorously attacked the program of privatization:

> Privatization can, eventually, be desirable. This would be the case for busi-
> nesses running deficits, those in bankruptcy, those which malfunction,
> or those that military dictatorship incorporated in the national heritage
> by shady business deals. It may also be the case within a national pro-
> gram of democratization of the capital of public enterprises through the
> sale of their shares to the employees and all the Brazilians who want to
> invest their savings in these businesses. But the privatization that is tak-
> ing place is nothing like this. On the contrary, what is happening is the

disposal of the national heritage, a heritage which is indispensable to the autonomous development of our economy, which would become irrevocable if it were privatized. (7)

Darcy will concede privatization if it is populist for Brazil, and Quayle wants royalties if it helps big businesses in the United States. Through a simple quid pro quo, both can get what they want. The Brazilians can pay royalties on American artificial information if, in turn, Americans pay royalties on Brazilian natural information. The terms of trades are such that on net, equal protection of artificial and natural information will be very favorable for Brazil. Of course, there will be winners and losers within both Brazil and the United States.

The following day, 9 August 1991, a follow-up story appeared in *Jornal do Brasil*: "Collor Criticizes Businessmen," 1, 3. Collor again takes up the issue of patents: "The issue of patents, trademarks, the question of intellectual property, of information reserves, all these points mark the entry of Brazil into the international market, obtaining her place of sovereignty and competitiveness in the international market." True enough. But inasmuch as Brazil is presently free riding on patents and other intellectual properties, its entry would mean an even greater drain of hard currency. But Collor has little choice. The U.S. ultimatum was reported five days later: "The U.S. Will Only Transfer Technology if Brazil Recognizes Patents," *Jornal do Brasil*, 13 August 1991, 3. President Bush's special envoy for science and technology, Allan Bromley, communicated the threat to Brazilian vice president Itamar Franco: "The USA will not negotiate any accord, if by November when the present term expires, Brazil has not yet implemented a law of patent protection. This was the first time the issue was put so bluntly."

13. On the rising preeminence of Japan in international finance, see Clyde H. Farnsworth, "IMF Panel Votes to Add $60 Billion to Pool for Loans," *New York Times*, 9 May 1990, A1, D6.

14. David Swinbanks and Alun Anderson,"Green Roots and Red Ends,"*New Statesman and Society*, 23 November 1990, 14–15; "Japan and Brazil Team Up," *Nature*, 9 March 1989, 103.

15. In economic theory, this game is known as a prisoner's dilemma. It is especially prevalent where competition among firms is stiff. There are today about 500 biotech firms worldwide. See Satish Jha, "Biotech and the Third World," *Economic and Political Weekly*, 9 June 1990, 1243–46. If firms attempt collusion, a prisoner's dilemma will emerge.

Chapter 9

1. On the mass appeal of debt-for-nature swaps, see "Forgive Debt, Finance Nature," *New York Times*, 16 July 1991, A18. In the Bolivian swap, see the

press release "Bolivia Sets Precedent with First Ever 'Debt for Nature Swap,'" *Conservation International*, 13 July 1987. The issue of policing a 1.5-million-hectare tropical forest was noticed early on by the economist Steis Hansen in "Debt-for-Nature Swaps," *Ecological Economics* 1 (1989): 77–93. Hansen argues: "What appears to hold true is that the debt-for-nature swap must take the form of compensation that can be stopped if the forest management agreement is not strictly adhered to. This suggests annual or other periodic compensation payments to make the country choose to abide by the agreement rather than to breach it" (83). The theory versus the reality of the swaps is made poignantly clear in Rhona Mahony, "Debt-for-Nature Swaps: Who Really Benefits?" *The Ecologist* 22, no. 3 (1992): 97–103. She concludes: "The only beneficiaries of debt-for-nature swaps at present are the Northern banks."

2. Genestead leases would be a good deal for Brazil. Twenty years ago, the government was giving away permanent titles to 100-hectare lots for as little as $700, payable over 20 years. See, for example, Alex Houmatoff, *The World Is Burning: Murder in the Rainforest* (New York: Avon Books, 1990), 51. Houmatoff quotes a typical cattle rancher: "If the international community is really worried about saving the Amazon they should send each family two or three thousand dollars a year, and we won't touch a leaf or raise a head of cattle" (311).

3. For the property-rights interpretation of share tenancy, see S.N.S. Cheung, *The Theory of Share Tenancy* (Chicago: University of Chicago Press, 1970). For a recent critique, see James Peach and Kenneth Nowotny, "Sharecropping Chicago Style: The Oppressed Landlord and the Inefficient Peasant," *Journal of Economic Issues* 26, no. 2 (1992): 365–72.

4. For an in-depth look at what can be extracted from the rain forest without risking extinction, see Annette Lees, *A Representative Protected Forests System for the Soloman Islands* (Mauria Society, P.O. Box 756, Nelson, New Zealand, 1991), and Anthony Anderson, ed., *Alternatives to Deforestation: Steps Toward Sustainable Use of the Amazon Rain Forest* (New York: Columbia University Press, 1990). Charles M. Peters of the Institute of Economic Botany at the New York Botanical Garden calculates the value of pasture in the Peruvian Amazon over a 50-year period as $2,960 per hectare and the value of the land for rubber, fruit, and other edible items as $6,330 per hectare. The calculation made the popular press in "Rain Forest Worth More If Uncut, Study Says," *New York Times*, 4 July 1989, 18.

5. During the next 15 years, NASA will propose expenditures of $30 to $40 billion for Earth Observing Systems (EOS) to monitor global warming, deforestation, and desertification. NASA will call this program Mission to Planet Earth. See Pamela Zurer, "NASA to Revamp Earth Observing System," *Chemical and Engineering News*, 14 October 1991, 20–21.

6. Harold Demsetz, "Toward a Theory of Property Rights," *American Economic Review*, no. 2 (1967): 367.

Chapter 10

1. On the cognitive dissonance of economists toward global warming, see Thomas Schelling, "Some Economics of Global Warming," *American Economic Review* 82, no. 1 (1992): 1–14. The opening sentence says it all: "Global warming from carbon dioxide was an esoteric topic 15 years ago, unknown to most of us" (1). Although global warming has been known in the scientific community for at least a century, heavily discussed over the past 30 years, and elaborately modeled for the last 10 years, "global warming" was not legitimate in the economic community until most of "us" (read: neoclassical economists) recognized it. The fact that the *American Economic Review (AER)* is ranked by U.S. university faculties as the leading journal of economics and places Schelling's article in the lead position indicates that its distinguished editorial board must also think global warming is something new and exciting.

2. Jeremy Leggett, ed., *Global Warming: The Greenpeace Report* (Oxford: Oxford University Press, 1990), 164; Jeremy Leggett, "Global Warming: A Greenpeace View," in *Global Warming*, ed. Leggett, 460.

3. Cognitive dissonance, like the greenhouse effect, is also not new! An early comprehensive treatment can be found in Leon Festinger, *A Theory of Cognitive Dissonance* (Stanford, Calif.: Stanford University Press, 1957). Nevertheless, it took a whole generation before cognitive dissonance would work its way into the *American Economic Review*: G. A. Akerlof and W. T. Dickens, "The Economic Consequences of Cognitive Dissonance," *AER* 72, no. 3 (1982): 307–19.

4. Gary Becker, "The Hot Air Inflating the Greenhouse Effect," *Business Week*, 17 June 1991, 9.

5. William D. Nordhaus, "Greenhouse Economics: Count Before You Leap," *The Economist*, 7 July 1990, 21–24.

6. On U.S. reluctance to acknowledge the greenhouse effect, see Allen L. Hammon, Eric Rodenburg, and William R. Moomaw, "Calculating National Accountability for Climate Change," *Environment* 33, no. 1 (1991): 11–35.

7. Although a minority opinion, the dissent over global warming also comes from learned climatologists such as Richard S. Lindzen, a professor at MIT. See "Greenhouse Skeptic Out in the Cold," *Science* 246 (1989): 1118.

8. Jeremy Rifkin, *Entropy: Into the Greenhouse World*, 2nd ed. (New York: Bantam Books, 1989).

9. When the AEA session "Economics and Entropy" opened to discussion, I commented that physicists would find it odd that economists would talk about entropy without ever mentioning a large body of literature known as nonequilibrium thermodynamics (NET), the pioneer of which, Ilya Prigogine, won a real Nobel Prize. See Ilya Prigogine, *From Being to Becoming* (San Francisco: Freeman, 1980), and *Order Out of Chaos* (New York: Bantam, 1987). I briefly stated some of the basic tenets of NET. To my surprise, all the subsequent com-

ments from the audience were in agreement with me and in disagreement with the panelists. This divergence between the audience and the panelists can also be explained in terms of NET. The organizer is a bifurcation point, and his choice of other panelists is an amplification effect. So, if the organizer believes entropy is just a metaphor, he chooses others with like beliefs. Unanimity obtains. The NET literature is now vast. The literature ranges from energy network diagrams—see Howard T. Odum, *Systems Ecology* (New York: Wiley, 1983)— to patterns of culture—see Richard N. Adams, *The Eighth Day: Social Evolution as the Self-Organization of Energy* (Austin: University of Texas Press, 1988). My previous work has also spanned this domain: Joseph H. Vogel, "Evolution as an Entropy Driven Process: An Economic Model," *Systems Research 5*, no. 4 (1988): 299–312; "Entrepreneurship, Evolution, and the Entropy Law," *Journal of Behavioral Economics* 18, no. 3 (1989): 185–204; "Uninvited Guests: A Thermodynamic Approach to Resource Allocation," *Prometheus* 9, no. 2 (1991): 332–45.

10. The concept of "bifurcation points" can be easily misunderstood. For example, in *Earth in the Balance: Ecology and the Human Spirit* (Boston: Houghton Mifflin, 1990), Al Gore likens the pending treaties for sustainable development in the 1990s to the missed opportunities to stop Nazi aggression in the 1930s. Gore identifies the bifurcation point of Nazi aggression as Kristallnacht— 9 November 1936—the bloody night when Hitler's thugs rampaged Jewish homes, businesses, and synagogues. Looking back, it is easy to see that Kristallnacht was indeed a bifurcation point. But at the time it was probably not so obvious. For example, is the 1992 skinhead attack on an immigrant asylum block in the Baltic port city of Rostock yet another Kristallnacht, with all its attendant horrors? Or is it an isolated incidence of violence and hate—a dead end? The concept of bifurcation points is that it is easy to trace the deterministic path ex post facto, but ex ante, there are so many conceivable bifurcation points that the path on which history will unfold is beyond our lens of resolution.

11. Arthur S. Eddington, *The Nature of the Physical World* (New York: Macmillan, 1929), 74.

12. Theodosius Dobzhansky, an evolutionist's evolutionist, once remarked: "Nothing in biology makes sense except in the light of evolution" (Foreword to Theodosius Dobzhansky, Francisco J. Ayala, G. Ledyard Stebbins, and James W. Valentine, *Evolution* [San Francisco: Freeman, 1977]). The quote originally appeared as the title of an article by Theodosius Dobzhansky in *American Biology Teacher* 35 (1973): 125–29. There is a minority school of economists who believe that something similar holds true for their discipline: nothing in economics makes sense except in the light of the entropy law. These economists emphasize the irreversible quality of all decisions as well as the predictable patterns of nonrational behavior that were once highly dissipative to a hunter-gatherer environment.

13. Douglas R. MacAyeal, "Irregular Oscillations of the West Antarctic Ice Sheet," *Nature*, 3 September 1992, 29–32.

14. On leaking methane, see Deborah MacKenzie, "Leaking Gas Mains Help to Warm the Globe," *New Scientist*, 22 September 1990, 24, and Max K. Wallis, "Leaky Answer to Greenhouse Gas," *Nature*, 1 March 1990, 25–26. For the carbon release attributed to tropical deforestation, see Norman Myers, "Tropical Forests," in *Global Warming*, ed. Leggett, 396.

15. Stephen H. Schneider, "The Science of Climate-Modelling and a Perspective on the Global-Warming Debate," in *Global Warming*, ed. Leggett, 44.

16. There are now hundreds, if not thousands, of excellent articles in the popular press regarding global warming and public policy. See, for example, Jon R. Luoma, "Gazing into Our Greenhouse Future," *Audubon*, March 1991, 52–59. For technical details on the policy options, see Joshua M. Epstein and Raj Gupta, *Controlling the Greenhouse Effect: Five Global Regimes Compared* (Washington, D.C.: Brookings Institution, 1990); Michael Grubb, *The Greenhouse Effect: Negotiating Targets* (London: Royal Institute of International Affairs, 1989); Michael Grubb et al., *Energy Policies and the Greenhouse Effect*, vol. 2: *Country Studies and Technical Options* (London: Royal Institute of International Affairs, 1991); Michael Grubb et al., *Emerging Energy Technologies: Impacts and Policy Implications* (London: Royal Institute of International Affairs, 1991); Terry L. Anderson and Donald R. Leal, *Free Market Environmentalism* (Boulder, Colo.: Westview Press, 1991); Peter R. Hartley and Michael G. Porter, "A Green Thumb for the Invisible Hand" (Manuscript, Tasman Institute, Melbourne, Australia, 1990); and "Special Issue: The Economic Costs of Reducing CO_2 Emissions," *OECD Economic Studies*, no. 19 (1992).

17. Even those economists who endorse an international carbon tax over property rights doubt the climatological evidence of global warming is conclusive. See, for example, point vi of the abstract to Anwar Shah and Bjorn Larsen, "Global Warming: Carbon Taxes and Developing Countries" (Paper prepared for presentation at the 1992 Annual Meetings of the American Economic Association, 3 January 1992, New Orleans):

> Tradeable permits represent a preferred alternative to carbon taxes in the event a critical threshold of the stock of carbon emissions beyond which temperatures would rise exponentially was known. Given our current lack of knowledge on the costs of carbon emissions reductions and the threshold effect, a carbon tax appears to be a superior and more flexible instrument to avoid potentially large unexpected costs.

18. Jos Haynes and Brian S. Fisher, "Economics and the Environment: The Greenhouse Effect. A Case Study" (Australian Bureau of Agricultural and Resource Economics, GPO Box 1563, Canberra 2601, paper CP89945), 8 (Paper presented at "Economics and the Environment," Economics Society, Victorian Branch, Conference, Melbourne, 21 June 1990). Haynes and Fisher reach a solution that many analysts are converging on: "One thing is certain: common property problems can only be solved through regulation of activity or through the

allocation of property rights. Either of these courses in relation to the green-house effect will require a revolution in attitudes by the nations of the world" (11).

19. Most environmentalists see global warming as yet another example of Garrett Hardin's "The Tragedy of the Commons," *Science*, 13 December 1968, 1243–48. Hardin explained a general mechanism in a specific instance. He described how pastures in the Middle Ages were degraded if left in the public domain; each peasant had an incentive to overgraze. Historically, property rights did in fact emerge in the form of "stinting"—that is, limiting "the pasturage of common land to a person. Late Middle English" (*The Shorter Oxford English Dictionary of Historical Principles*, 3rd ed.). See Carl Dahlman, *The Open Field System and Beyond* (Cambridge: Cambridge University Press, 1980).

20. Stephen Boyden, Stephan Dovers, and Mega Shirlow, *Our Biosphere Under Threat: Ecological Realities and Australia's Opportunities* (Melbourne: Oxford University Press, 1990), 185. For an update on Australian scientific efforts to address global warming, see Office of the Chief Scientist, Department of the Prime Minister and Cabinet, *Scientific Aspects of Major Environmental Issues: Climate Change* (Canberra: Australian Government Publishing Service, 1992).

21. Permits for greenhouse-gas emissions will work well if there is near unanimity and the recalcitrant country is small and trade dependent. The horror is the United States. Its greenhouse contribution is roughly equal to the global target, 20 percent. Any trade sanctions would be pointless, inasmuch as the United States could be self-sufficient.

22. Grubb, *Greenhouse Effect*, 40. The fairness issue does not escape the critics of privatization. For example, Fred Pearce expresses trepidation over bids for a greenhouse auction in "Bids for the Greenhouse Auction," *New Scientist*, 4 August 1990, 47–48. Pearce writes:

> For many environmentalists, international trade is the cause of many of the world's environmental problems and an unlikely mechanism for solving them. And why should an individual "right to pollute" be in the hands of the government? I confess to sharing this unease. But . . . there is an urgent need to find new rules by which the world can order its affairs. If tradeable permits will help to persuade the ideologues of the White House that they should help to save the planet's climate, then all power to them. (48)

When it comes to fairness over the initial distribution of property rights, see Anthony Scott, "Property Rights and Property Wrongs," *Canadian Journal of Economics* 16, no. 4 (1983): 555–73. Scott offers this warning:

> [I]f you think existing rights to land and natural resources lead resource users to behave in a wasteful or violent fashion, try to repair or recon-struct their rights. It is true that new real rights systems may have high transactions costs, and you will have to think about these. And it is true

that you cannot set up such a system without making a permanent distributional judgment, similar to that made by William [the Conqueror] when he gave his officers certain rights to all of England. But alternative remedies will have their own transactions costs and distributional consequences. If you can be satisfied on these two problems, you can proceed. (568)

23. Thomas Lovejoy, coeditor with Robert L. Peters of *Global Warming and Biological Diversity* (New Haven, Conn.: Yale University Press, 1992), expresses the obvious well: "I fail to see that there's any conclusion to draw from all of this other than that there will be massive extinction no matter what we do in the way of conservation. Therefore, the only logical conclusion is to prevent as much of the climate change as possible." See William K. Stevens, "Global Warming Threatens to Undo Decades of Conservation Efforts," *New York Times*, 25 February 1992, B8.

Chapter 11

1. See Al Eichner, ed., *Why Economics Is Not Yet a Science* (Armonk, N.Y.: Sharp, 1983); "The Lack of Progress in Economics," *Nature*, 7 February 1985, 427–29; and the response by Partha Dasgupta and Frank Hahn, "To the Defence of Economics," *Nature*, 17 October 1985, 589–90. Admittedly, Eichner did ignore and confuse much significant neoclassical literature, as Dasgupta and Hahn note. But it is they who ignore the central message of Eichner's work: assumptions first must be validated before economists can proceed with deductive reasoning. The steadfast defense of a failed neoclassical model has exasperated otherwise patient scientists who work at the interface of economics and conservation biology. For example, E. O. Wilson states flatly that "contemporary economics is bankrupt" (personal communication, 24 May 1993).

2. The latest statistic on starvation was made by the head of the United Nations Food and Agricultural Organization, Edouard Saouma. The figure was put at 780 million. See Pamela Bone, "World Hunger Is Scandal: UN Chief," *The Age*, 24 June 1993, 10.

3. The sociology of the economics profession is not lost on NET advocate Peter Allen: "The equilibrium hypothesis is tenacious, mainly because it avoids all the real difficulties of life, and can lead to elegant theorems and lemmas, which are the very stuff of Ph.D.'s, professorial appointments and honorary degrees. Despite the fact that it flies in the face of everyday experience, it has therefore been the foundation on which the whole edifice of economic theory has been built" ("Evolution, Innovation, and Economics," in *Technical Change and Economic Theory*, ed. Giovanni Dosi et al. [New York: Pinter, 1988], 98).

4. Much of the criticism of neoclassical economics is as old as neoclassical economics itself! For a history of the questionable foundations on which the neo-

classical model was built, see Philip Mirowski, *Against Mechanism: Protecting Economics from Science* (Totowa, N.J.: Rowman & Littlefield, 1988).

The relaxation of diminishing returns, just one of the assumptions of the neoclassical model, results in exactly opposite policy prescriptions to the usual neoclassical recommendations. See W. Brian Arthur, "Positive Feedbacks in the Economy," *Scientific American*, February 1990, 80–84. Relaxing the other highly unrealistic assumptions transforms mainstream practice into what critic Charles A. S. Hall sees as a fraudulent exercise in "Sanctioning Resource Depletion: Economic Development and Neo-Classical Economics," *The Ecologist* 20, no. 3 (1990): 99–104. Regarding the irrational assumption of rationality in neoclassical economics, see Amartya Sen, "Rational Fools: A Critique of the Behavioral Foundations of Economic Theory," *Philosophy and Public Affairs* 6, no. 4 (1977): 317–44.

5. For an overview of some of the cross-fertilization of disciplines and the emergent fuzziness of economic boundaries, see P. Paolo Saviotti and J. Stanley Metcalfe, eds., *Evolutionary Theories of Economic and Technological Change: Present Status and Future Prospects* (Philadelphia: Harwood, 1991).

6. The late Kenneth Boulding advocated rights over reproduction many years ago. Orthodox economists were uninterested. See Kenneth Boulding, *The Meaning of the 20th Century: The Great Transition* (New York: Harper & Row, 1964), 121–36, and David Heer, "Marketable Licenses for Babies: Boulding's Proposal Revisited," *Social Biology* 22, no. 1 (1975): 1–16. In the rhetoric of this book, Boulding's licenses for babies is the creation and attenuation of the right to reproduce.

7. Environmental ethics can become bizarre. For example, Michael Kennedy throws down the gauntlet in *Australia's Endangered Species* (Melbourne: Simon and Schuster, 1990): "First, it must be recognized that all species on this planet maintain an inalienable right to existence" (17). Why must it be recognized? In light of NET, Kennedy's dictate sounds a bit strange: Why should a sequence of base pairs have any rights whatsoever?

8. Quoted in Jeremy Legett, ed., *Global Warming: The Greenpeace Report* (Oxford: Oxford University Press, 1990), 113.

9. The Brundtland Report, contained in World Commission on Environment and Development, *Our Common Future* (New York: Oxford University Press, 1987), is often bandied about in any discussion of the environment. Among its goals is the application of laws that favor sustainable development. However, from thermodynamics we know that time only gives the illusion of sustainability—nothing is sustainable in the long-run. Grumblings over the word "sustainability" are coming from many quarters. Murray Gell-Man, a Nobel laureate in physics, berates it in "Visions of a Sustainable World," *Engineering and Science*, Spring 1992, 5–10. Terry L. Anderson and Donald R. Leal also perceived its flaws for economics in *Free Market Environmentalism* (Boulder, Colo.: Westview Press, 1991), 167–72.

10. Perhaps the most famous adaptive ethic is the Marxist aphorism "To each according to his needs, from each according to his ability." The ethic makes complete adaptive sense for a hunter-gatherer economy. In such an economy, needs are readily visible in terms of food, shelter, and protection; ability is readily visible in terms of sex, age, and health. But today this ethic is maladaptive. How do we judge needs and abilities in a postindustrial society? We cannot. Nevertheless, the rhetoric will still sound good because our preference for ethics has not evolved beyond that of a hunter-gatherer.

11. Convention on Biological Diversity (Draft distributed by the United Nations Environment Programme at RIO'92, June 1992, Rio de Janeiro), republished in *RECIEL*, 17 December 1992, 339–69. [Italics added]

12. Clinton's interpretation of the convention integrates with a bipartisan pattern of abuse by American presidents to interpret treaties to mean anything they wish—no matter how contrary to the letter of the law. See "Bioconvention Ratification Waits as Nations Draft Interpretative Statements," *DIVERSITY* 1, 2 (1993): 8–10. The last such outlandish abuse was during the Reagan administration. Reagan's legal adviser, Abraham Sofaer, reinterpreted the 1972 Anti-Ballistic Missile (ABM) treaty to allow testing and deployment of ballistic missiles. What drove the reinterpretation of the ABM treaty was special interests—hundreds of billions of dollars extracted from the American public through the Strategic Defense Initiative. See Tim Weiner, "Lies and Rigged 'Stars Wars' Test Fooled the Kremlin, and Congress," *New York Times*, 18 August 1993, A1, A15. What drives the interpretation of Article 16 of the Convention on Biological Diversity is special interests—hundreds of billions of dollars extracted from the Third World through the concept of the "common heritage of mankind." As reported in the *New York Times* immediately following the President's remarks on Earth Day 1993: "G. Kirk Raab, president of Genentech Inc., which was part of the coalition that developed the interpretative statement said, 'This is a major step forward for both businesses and environmentalists'" (Richard L. Berke, "Clinton Supports Two Major Steps for Environment," *New York Times*, 22 April 1993, A1, A10).

13. In consultation with patent attorneys, the CEO of Genentech, G. Kirk Raab, dashed off a quick note to President Bush just before RIO'92. The message: nix the Convention on Biological Diversity. Somehow the message leaked out to the scientific staff of Genentech and caused an uproar. Raab defended the note to *Nature* with these remarks: "I don't believe mixing in industrial property rights is the least bit appropriate. If you dig up a little piece of dirt in Naples . . . or pick a flower in Ecuador, I don't think there is necessarily a requirement that the country has some predetermined economic rights" (Sally Lehrman, "Genentech Stance on Biodiversity Riles Staff," *Nature*, 9 July 1992, 97). Similar letters went out to Bush from the Association of Biotechnology Companies and the Pharmaceuticals Manufacturer's Association.

14. While we dispense with the expression "sustainable development," we should

also get rid of "harmless use," its legal cousin. "Harmless use" makes no entropic sense. It gives people the false impression of equilibrium when the world is in a perpetual nonequilibrium. Everything destroys or is destroyed. It is just a matter of degree or time.

15. Regarding the failures of the World Bank, see Pratap Chatterjee, "World Bank: A Loan Voice Concedes Its Failures," *Sunday Age*, 8 November 1992, World Section, 2.

16. Learning to say *no* can occur incredibly quickly, as reported in Nicholas Schoun, "The Spirit of Rio in Trouble," *The Age*, 7 June 1993: "In a blow to global solidarity, several wealthy donor nations announced large cuts in their Third World aid budgets soon after the Summit" (8).

17. Quoted in Marshall Sahlins, *The Use and Abuse of Biology: An Anthropological Critique of Sociobiology* (Ann Arbor: University of Michigan Press, 1976), 10.

18. Neither the Pastry War of 1838 nor the Football War of 1969 was fought over something as inconsequential as a croissant or a goal. France wanted compensation for expropriations of its citizens' property in Mexico; El Salvador wanted claim to lands in Hondorus on which Salvadoreans had squatted. Nevertheless, both the croissant and the goal triggered a battle over property rights.

19. Statistics in Brazil have become surreal. Some claim that there are 3 million children in the streets committing petty and not so petty crimes. The urban middle class is truly terrified of the youthful underclass. "The civil and military police, private security guards for shops and supermarkets, commissioners of minors, justice officials and judges are involved directly or indirectly with the extermination of children." This was the conclusion of the final 100-page report of the Comissão Parlamentar de Inquérito (CPI) do Menor (Parliamentary Commission Inquest on Minors by the Legislative Assembly). See "CPI Gives Names of Those Who Kill Minors in Rio," *Jornal do Brasil*, 2 November 1991, 14. And the violence is not confined to children or to urban areas. See *Brazil: Authorized Violence in Rural Areas* (London: Amnesty International Publications, 1989).

20. Army general Thaumaturgo Sotero Vaz, Chief Military Command for the Amazon, publicly expressed fears of invasion on the pretext of human-rights abuse. Indeed, both the massacre of Yanomami Indians and the extermination of street children in July 1993 were widely reported in the United States. See, for example, James Brooke, "Questions Raised in Amazon Killings," *New York Times*, 29 August 1993, 14, and "8 More Children Reported Killed in Brazil," *New York Times*, 29 August 1993, 14. The fears of U.S. intervention are heightened by a $600 million radar project located across the borders in Colombia and Guyana. See James Brooke, "Brazil's Army Casts U.S. as Amazon Villain," *New York Times*, 19 August 1993, A6. Politicians in the Amazon do not dismiss the threat. In an interview, the governor of the Amazons, Gilberto Mestrinho, states: "If there is foreign interference here, this will be the next Viet Nam" (*Playboy* [Brazilian ed.], February 1992, 27–38).

21. The Gulf War was foreshadowed by Jimmy Carter: "An attempt by any outside force to gain control of the Persian Gulf Region will be regarded as an assault on the vital interests of the United States of America, and such an assault will be repelled by any means necessary, including military force" ("State of the Union Address," *Department of State Bulletin*, no. 2035 [1980]: 38A–38D), quoted in Alexander A. Arbatov, "Oil as a Factor in Strategic Policy and Action: Past and Present," in *Global Resources and International Conflict: Environmental Factors in Policy and Action*, ed. Arthur H. Westing (New York: Oxford University Press, 1986), 26.

22. Regarding Kodiak Island, see "Born Free, Sold Dear: America's Premier Bear Refuge Is Up for Bids Because Native Alaskans Have Run Out of Money," *Newsweek*, 6 May 1991, 52–54.

Glossary

ATTENUATION The constraint of a legal property right.

BIFURCATION POINT A juncture at which future development is channeled by the path chosen. Because there are so many junctures at any one moment, and so many moments over time, the future becomes a thick tangle of possible paths. Prediction becomes impossible. Nevertheless, looking back, one can trace the causation to critical junctures. *See* BOUNDARY CONDITIONS and ENTROPY LAW.

BIODIVERSITY Usually left undefined, but taken to mean species diversity.

BIOTA All living things.

BIOTECHNOLOGY The application of scientific and engineering principles to the processing of materials by biological agents to provide goods and services.

BOUNDARY CONDITIONS The state of the physical environment at the moment of an event. *See* BIFURCATION POINT and ENTROPY LAW.

CETERIS PARIBUS Latin for "holding all things equal." Expresses the analytical notion of isolating the effects of one variable when causation is multivariate.

COMMONS Assets controlled by a specified group.

CULTURALLY CODED FUNCTIONS (CCFs) Those services rendered from the manifestation of cultures. *See* GENETICALLY CODED FUNCTIONS (GCFs).

DECONSTRUCTION A theory of literary criticism that denies the absoluteness of any meaning assigned to a text.

DE FACTO Description of a state of affairs that is illegal or illegitimate but, for practical purposes, must be recognized as existing. *See* DE JURE.

DE JURE   Description of a state of affairs that is legal and legitimate but, for practical purposes, may or may not exist. *See* DE FACTO.

EDGE EFFECT   The erosion of a virgin habitat at its border with a clearing.

EFFICIENCY   The measure of output to input. *See* EQUITY.

ELASTICITY   The percentage change in one variable with respect to a percentage change in another variable. For example, the price elasticity of the quantity demanded, e, is given by the formula

$$e = \Delta Q \cdot P / \Delta P \cdot Q$$

where Δ is change; Q, quantity; and P, price.

ENTROPY LAW   Also known as the second law of thermodynamics. By the first law, energy can be neither created nor destroyed; by the second law, although neither created nor destroyed, energy and matter go from states of higher order to those of lower order or, alternatively, from states of available work to states of unavailable work. *See* BIFURCATION POINT and BOUNDARY CONDITIONS.

EQUITY   The measure of the distribution of output. *See* EFFICIENCY.

EVOLUTION   A change in gene frequencies.

EXTINCTION   The loss of the last member of a species.

FREE RIDING   The enjoyment of benefits of a good or service without bearing any of the costs of production of that good or service. Occurs because exclusion from consumption is prohibitively costly due to the nature of the good or service.

GENESTEAD   The ceding of a government title to genetic information in exchange for habitat management over a specified period of time. *See* HOMESTEAD.

GENETICALLY CODED FUNCTIONS (GCFs)   Those services rendered from the instructions of DNA. *See* CULTURALLY CODED FUNCTIONS (CCFs).

GEOGRAPHIC INFORMATION SYSTEM (GIS)   Software packages that enable data entry and management for physical–spatial locations.

GLOBAL WARMING   The trapping of solar radiation by a variety of gases, such as CO_2, CH_4, NO_x, and H_2O.

GREENHOUSE EFFECT   *See* GLOBAL WARMING.

HOMESTEAD   The ceding of a government title to land in exchange for land improvements over a specified period of time. *See* GENESTEAD.

INFORMATION THEORY   A body of knowledge built on the notion that phenomena can be reduced to probabilities and quantified into bits of information. The quantification is via the Shannon–Weaver equation

$$H = - \Sigma P_i log P_i$$

where H is the bits; P_i, the probability of the occurrence; and *log*, the logarithmic function.

INTELLECTUAL-PROPERTY RIGHT   The legal recognition of the products of intellect. The rights vary according to the type of product—patents over pharmaceuticals, copyrights over books, trademarks over designs, trade secrets over customer lists, and sui generis legislation over semiconductors and plant varieties.

LANDRACES   Populations developed from artificial selection in agriculture done over many years.

MADISON AVENUE   A street in New York City where American advertising agencies are heavily concentrated. Over the years, the name of the street has become a metaphor for the psychological manipulation of the public through advertising.

NEOCLASSICAL ECONOMICS   Also known as orthodox or mainstream economics. A methodology that conceptualizes the economy as driven by rational utilitarian agents making reversible decisions.

NONEQUILIBRIUM THERMODYNAMICS (NET)   The field of enquiry that looks at open systems that exchange energy and matter with their surroundings. Order emerges through facilitating disorder. *See* BIFURCATION POINT and ENTROPY LAW.

NONGOVERNMENTAL ORGANIZATIONS (NGOs)   Organizations that are not financed by government directly or indirectly. However, many organizations identify themselves as nongovernmental, but are funded indirectly through government-sponsored industries or grants.

NONINDIGENOUS SPECIES   An interbreeding group of organisms that have been introduced into a habitat in which they did not evolve. The possibility exists for an exponential expansion of the organism and the extermination of native species due to predation or the crowding out of resources.

NORTH   The group of countries that have achieved economic development. Usually, a country of the North is industrial and is located in the Northern Hemisphere, while a country of the South is agricultural and is located in the Southern Hemisphere. But not always. For example, New Zealand is agricultural and located in the Southern Hemisphere, but its economic development puts it in the North.

ORIGINAL INTENT   The legal theory that maintains that (1) the reasoning embodied in the Constitution can be discovered, and (2) once discovered, that reasoning should be extended to new circumstances.

PAPER PARKS   Reserves that are demarcated on maps but fail to exist in reality.

PROPERTY RIGHT (ECONOMIC)   The ability to control, dispose, and derive benefits from a process or resource. *See* PROPERTY RIGHT (LEGAL).

PROPERTY RIGHT (LEGAL)   The recognition by the state of one's ability to control, dispose, and derive benefits from a process or resource. *See* PROPERTY RIGHT (ECONOMIC).

PROXIMATE CAUSATION   *See* ULTIMATE CAUSATION.

RESIDUAL CLAIMANT   The person who derives benefits from an asset regardless of whether or not he holds legal title. *See* PROPERTY RIGHT (ECONOMIC) and PROPERTY RIGHT (LEGAL).

SHANNON–WEAVER EQUATION   *See* INFORMATION THEORY.

SOUTH   *See* NORTH.

SPECIES   A group of interbreeding organisms.

SQUATTER   Someone who takes possession of a property without legal right.

SUSTAINABILITY   The ability to extract services from a habitat without changing the habitat.

TRANSACTION COST   The costs of monitoring, policing, negotiation, and communication entailed in production and exchange.

ULTIMATE CAUSATION   The "why" question in contrast to proximate causation, the "how" question. The distinction can be illustrated by the human hand. Anatomy and physiology explain *how* the hand works —the precision grip and the opposable thumb. Evolution explains *why* we have a precision grip and an opposable thumb.

Index

Abortion, 47, 49
AIDS, 67–68
Altieri, Miguel, 12
Alvarez, Walter, 92
Amazon, 35, 40, 44–45, 63, 73–74, 80, 85–86, 99, 107, 111, 113
American Seed Trade Association, 68
Amerindian, 26, 44, 85, 111, 113
Amplification effects, 92–93, 95, 111
Analogy, 16, 23, 29–31, 36, 55–56, 94
Anderson, Terry, 97
Ants, 18–19
Argentina, 102, 107, 112
Attenuation, 79, 81, 104–6, 109
Australia, 21, 23–24, 54, 65–66, 80, 100
AZT, 67–68, 86

Batrachian, 11, 50
Becker, Gary, 90
Beetles, 82
Bifurcation, 92–94, 111
Biodiversity, 12–13, 15, 17–18, 20–21, 25, 41, 60, 67–68, 74–75, 80, 91, 102–4, 106–9
Biota, 18, 20–21, 34, 57
Biotechnology, 26, 108
Boltzmann equation, 16
Bork, Robert, 70
Boundary conditions, 92–93, 111. See also Thermodynamics
BR-364, 45, 74. See also Amazon
Brady, Nyle C., 12

Brazil, 35, 44–45, 72–75, 77, 83, 85, 99, 107, 111, 113
Brown, Lester, 9, 12
Burley, F. William, 12
Burroughs Wellcome, 67–68
Bush, branching, 15, 80
Bush, George, 28, 72–73, 85, 98, 108

Cairns, John, 12
Carbon dioxide, 91, 95–96, 99. See also Greenhouse gas
Carbon tax, 100
Carlin, Norman, 60
Catharanthus roseus, 66. See also Periwinkle, rosy
Centralization, 56, 113
Chalker, Lynda, 112
Chemotherapeutics, 11, 37, 42, 67, 72
Chlorofluorocarbons, 96. See also Greenhouse gas
Classification, 44–46, 50–51, 57, 59, 61–62. See also Taxonomy
Climatology. See Global warming; Greenhouse gas
Clinton, Bill, 94, 109
Coase, Ronald, 60
Cognitive dissonance, 89–90, 96
Commons, 26, 53, 55–56, 59, 63, 80
Communication, 56, 58, 85
Complexity, 31, 104, 114

Conservation, 3–6, 9–13, 15–18, 20–21, 28–34, 40–41, 44–46, 50, 54, 56–57, 61, 66, 68–69, 82, 84, 92, 101–2, 104–10, 112, 114
Constitution, U.S., 24, 68–70, 72–74
Copyright, 4, 25, 29–31, 36, 38, 64, 79. *See also* Property right: intellectual
Cost–benefit analysis, 12, 15, 22, 106
Culturally coded function (CCF), 41–43

Darwin, Charles, 15
Dasgupta, Partha, 103
Deconstruction, 70
Delineation, 25–29, 52–53, 55–57, 59, 63, 65, 82–83
Demsetz, Harold, 25, 26, 86
Dinosaur, 16, 92
Disney World, 107
Dissipative structure, 93, 106
DNA sequencing, 63
Dolphin, 9
Doomsday, 101
Duckbill platypus, 21
Duke University, 82

Earth Resources Research, 97
Easy riding, 29–30
Eddington, Sir Arthur, 93
Efficiency criterion, 5, 36, 43, 48, 59, 98, 113
Ehrenfeld, David, 9, 12, 21
Ehrlich, Paul and Anne, 9, 12, 49, 60
Eichner, Al, 103
Elasticity, 40, 47, 66–68
Elephant, 11
Eli Lilly, 42
Entropy law, 92–93, 96. *See also* Thermodynamics: second law of
Environmental Protection Agency (U.S.), 28, 108
Epstein, Joshua, 97

Equity, 87, 98–100, 106, 110
Evolution, 15–17, 20, 22, 40, 41, 72, 80, 103, 106
Extinction, 3, 5–10, 12–13, 15, 20, 25, 31, 33–34, 36, 45, 50, 62, 69–70, 81, 85, 88, 101, 105, 107, 111, 113

Family planning, 46–47, 49, 111
Faux ami, 23–24
Feminism, 47, 49
Fisher, Brian, 97, 100
Flatulence, 99
Football, 23–24, 111
Fossil fuels, 91, 95, 99–101
Free riding, 29–31, 56, 72, 108
Friedman, Milton, 54, 60
Fromm, Erich, 90

General environmental fund (GEF), 59, 110
Genestead, 5, 39–40, 43–49, 50–51, 58, 60–61, 63, 65, 74, 76, 78–79, 81, 84–87
Genetically coded function (GCF), 15, 17–22, 29–31, 38, 42, 54–55, 75, 86
Geographic information systems (GIS), 58
Global Biodiversity Strategy, 106–7, 109–10
Global warming, 5–6, 88–90, 94–97, 99, 101–2
Goodall, Jane, 41, 43
Goodland, Richard, 12
Greenhouse gas, 6, 85, 89–90, 92, 94–102, 105. *See also* Global warming
Grubb, Michael, 97–98, 100
Gupta, Raj, 97

Hahn, Frank, 103
Halons, 96. *See also* Greenhouse gas
Hamilton, William, 41

Hartley, Peter, 97
Hawaii, 50
Haynes, Jos, 97, 100
HIV, 67. *See also* AIDS
Homestead, 43–45
Human rights, 43, 62, 69, 83, 85, 109,
    111–14
Huntley, Brian, 11
Hurd, Douglas, 112

Ice age, 80
Iguaçu, Brazil, 107
Incentive, 25, 31, 33–34, 36, 45–48,
    50–51, 52, 61, 65, 71–72, 79, 82,
    84, 96, 98–100, 102, 114
Indigenous people, 4, 41, 75, 114. *See
    also* Amerindian
Information, artificial and natural, 38,
    65, 68, 70–72, 74–75, 79, 110, 114
  genetic, 4–6, 17, 20, 22, 25–26,
    29–31, 33, 36–38, 40–41, 43,
    45–47, 50, 53–56, 59, 63, 65–67,
    69, 71–72, 74–77, 79–80, 86–88,
    99, 101–4, 109, 114
  theory of, 16, 70–71
Intelligence, artificial, 59, 61, 110
InterAmerican Development Bank
    (IADB), 74
Intergovernmental Panel on Climate
    Change (IPCC), 90
International Monetary Fund (IMF), 4,
    73
International Union for the
    Conservation of Nature (IUCN),
    108
Inventory, biological, 38–39, 55, 57

Jackson, Michael, 37
Japan, 19, 74–75, 98
Justice, 4

Kahn, Herman, 101
Keynes, John Maynard, 78

Kimura, Motoo, 16–17
Kirkpatrick, Jeane J., 112
Kissinger, Henry, 112
Kodiak Island, 114
Kohl, Helmut, 94
Kuwait, 69

Language, 14–15, 30, 47, 49, 70–71,
    83, 86, 104
Leal, Donald, 97
Leggett, Jeremy, 89–90
Leopold, Aldo, 9
Liberation theology, 70, 83
Linnaeus (computer program), 61–
    62
Linnaeus, Carolus (Carl von Linné),
    57
Lovejoy, Thomas, 12

Madagascar, 11, 42–43, 67, 80
Madison Avenue, 10–11, 67
Madonna (Madonna Louise Ciccone),
    37
Maize, 19
Major, John, 94
Marshall, S. A., 57
Marxism, 83
McCloskey, Donald, 3
Mendes, Chico, 74, 83–84
Menem, Carlos, 102
Merrick, Laura C., 12
Mestrinho, Gilberto, 73
Metaphor, 15, 56, 85, 93, 102
Methane, 94–97, 99. *See also*
    Greenhouse gas
Mitterrand, François, 113
Monetarism, 77
Monotreme, 21
Moore, John, 18–20, 53
Myers, Norman, 9, 37

National Forum on BioDiversity, 8,
    13, 17

National Wildlife Refuge, 114
Neo-Darwinian synthesis, 15. *See also*
    Evolution
New York Rainforest Alliance, 8
Nitrous oxide, 96. *See also*
    Greenhouse gas
Nongovernmental organization
    (NGO), 8–9, 21, 109
Nonrational behavior, 93. *See also*
    Cognitive dissonance
Nordhaus, William, 90–91, 93, 97
Norplant, 111. *See also* Family
    planning
North, 3–4, 6–7, 10–11, 28, 32, 37,
    47, 49, 54, 61–62, 66, 69, 72–
    73, 75, 77, 79, 81, 84, 86–87,
    100, 108, 111–14. *See also*
    South

Ocana, Gilberto, 12
Original intent, 70–72
Owl, spotted, 9

Paper park, 28, 84, 108
Pastry War, 111
Patent, 4, 19, 25, 36–38, 61, 64–65,
    68–69, 71, 74, 79, 86–87. *See also*
    Property right: intellectual
Periwinkle, rosy, 11, 42, 50, 66–67,
    71–72
Permafrost, 95
Physician's Desk Reference (PDR),
    42
Poaching, 33, 84
Population, 5, 9, 12, 17, 21, 34, 38, 46,
    49, 87, 95, 100, 107
Porter, Michael, 97
Privatization, 3–4, 6, 10, 14, 17, 20, 23,
    29, 32, 34, 37, 42–45, 47–48, 50,
    54–57, 60, 66–69, 72, 75–76, 81,
    83–84, 88, 92, 99, 101–2, 104–6,
    113–14. *See also* Property right

Property right, 5, 13, 22–34, 38, 41–
    43, 46, 48, 52–53, 59–60, 64–65,
    72, 74, 79, 83–84, 86, 101, 104,
    110, 112, 114. *See also*
    Privatization
    intellectual, 4, 31, 38, 64–66, 105,
    107–8

Quayle, Dan, 72–73
Quinine, 75

Race, 14, 18, 20, 53, 55, 59, 63. *See*
    also Species: subspecies
Randall, Alan, 12
Ray, G. Carleton, 18
Refugia, 80
Residual claimant, 27, 33, 40, 77
Rhetoric, 3, 6, 8–12, 17, 20–21, 28,
    32, 34, 44, 56, 66–69, 72, 75, 85,
    94, 98, 104–6, 113–14
Rifkin, Jeremy, 93
Robinson, Michael H., 12
Royalty, 5, 33, 37–40, 42–44, 46–47,
    50–51, 55, 59–63, 65–68, 76, 79,
    80–81, 84, 86–88, 99, 109, 111,
    114

Saudi Arabia, 69, 112
Schneider, Stephen H., 97
Self-determination, 113
Semiotics, 70
Shaman, 42–43
Shannon–Weaver equation, 16, 71
Share tenancy, 5, 77, 79, 81
Shen, Sinyan, 12
Smith, Adam, 7
Software, 29–31, 36, 55
Soulé, Michael, 10
South, 3–7, 28, 32, 37–38, 41, 47, 50,
    54, 61–62, 66, 69, 72–76, 78–79,
    84, 86, 100, 108, 111. *See also*
    North

Species, 5, 9–11, 13–14, 16–22, 26, 34–35, 37–39, 41–43, 50, 53, 55, 57–59, 61–63, 66, 76, 80, 82, 85–86, 92, 101–2, 107, 110
  nonindigenous, 49–50, 78–79
  subspecies, 18, 61
Squatter, 28, 34, 46–47, 78, 85
Supreme Court, U.S., 70–72
Sustainability, 81, 104, 106–11, 113

Tapper, of rubber, 81, 83–84
TAXACOM (electronic bulletin board), 62
Taxonomy, 19, 31, 50, 57–63, 82, 110
Thailand, 54

Thermodynamics, 16, 92–94
  nonequilibrium (NET), 6, 104, 106, 111, 113
  second law of, 92–94
Third World debt, 4, 37, 69, 73, 77–78, 84, 111, 112
Trademark, 4, 38, 64, 79. *See also* Property right: intellectual
Trade secret, 4, 38, 42–43, 64, 79. *See also* Property right: intellectual
Transaction costs, 26–32, 47, 53–56, 65, 69, 72, 79, 83, 86, 104

Wetland, 11, 107
Wilson, E. O., 9, 13, 17, 41, 60
World Bank, 8, 33–34, 54, 73, 109
World Resource Institute, 108

Zea diploperennis, 19–20, 53